LIFE AND TIMES OF A BIG RIVER

Life and Times of a
BIG RIVER

AN UNCOMMON NATURAL HISTORY OF ALASKA'S UPPER YUKON

PETER J. MARCHAND

University of Alaska Press FAIRBANKS

University of Alaska Press
P.O. Box 756240
Fairbanks, AK 99775-6240

ISBN 978-1-60223-247-1 (paperback); ISBN 978-1-60223-248-8 (electronic)

Library of Congress Cataloging-in-Publication Data
Marchand, Peter J.
Life and times of a big river : an uncommon natural history
of Alaska's Upper Yukon / Peter J. Marchand.
pages cm
Includes bibliographical references and index.
ISBN 978-1-60223-247-1 (paperback) — ISBN 978-1-60223-248-8 (electronic)
1. Natural history—Yukon River Valley (Yukon and Alaska) 2. Biodiversity—Yukon River
Valley (Yukon and Alaska) 3. Scientifice expeditions—Yukon River Valley (Yukon and
Alaska). 4. Indians of North America--Yukon River Valley (Yukon and Alaska) 5. Yukon-
Charley Rivers National Preserve (Alaska)—History. 6. Marchand, Peter J.—Travel—
Yukon River Valley (Yukon and Alaska) 7. Biologists—Alaska—Biography. I. Title.
QH104.5.Y85M37 2015
508.798'6—dc23

Interior photos by Peter Marchand unless otherwise noted
Frontispiece photo of the author by Ed Holsten
Cover design by Dixon Jones
Cover photo by Brian Heaphy (brianheaphy.com)
Back cover photo by Garrett Clough

This publication was printed on acid-free paper that meets the minimum requirements for
ANSI / NISO Z39.48‚Äì1992 (R2002) (Permanence of Paper for Printed Library Materials).

CONTENTS

VI. HEADWATERS

VII. THE YEARS AFTER

PREFACE

It would be more than a little presumptuous to suggest this work has anything in common with Ernest Hemingway's, but Hemingway *did* teach me two important lessons in his writing of *A Moveable Feast*. He showed me that not only was it permissible to write about a period in one's life decades later, but suggested that it was more effective to write from a distance—to record his impressions about Paris while living in Idaho (and Cuba and Spain) and to write about life in Michigan while residing in Paris.

So it was that in 1997, while living in the deserts of Arizona, I began writing about a Yukon River expedition that I had participated in twenty-two years earlier. It would be another fifteen years before I finished the project, penning the last details from the mountains of Colorado.

In undertaking this drawn-out endeavor, however, I had advantages that Hemingway did not. I had in my possession both the field journals of two other men who accompanied me on that expedition and an extensive collection of photographs from the time. These kept the record straight and helped fill in details that memory might have lost in the intervening years. I also had fingertip access, via the internet, to important documents of the time, recording the exact events, for example, leading to the signing of the Alaska Native Claims Settlement Act that paved the way for our expedition.

The idea of a narrative about interior Alaska floated in and out of my mind in the years shortly after the 1975 expedition, but the first real motivation to write arrived unexpectedly in the mail during the winter of 1989, in a package from someone I didn't know. The package contained the field notes of Garrett Clough, my mammalogist colleague on the river, explaining rather matter-of-factly that upon Garrett's untimely death, his collections and papers had been given to Suffolk University, and the sender, one Robert J. Howe, thought I might like to have the Yukon-Charley field notes. I was disturbed by the news of Garrett's death, of course, but I was pleased that someone had gone through the trouble to find me. Winter of 1989 was a difficult time in my life, however, so I carefully packed the letter and notes away for another day. They resurfaced again eight years later, and that's when I started this endeavor in earnest. The project would get sidelined three more times for other books, but eventually persistence won out—the story all the richer, I hope, for its long incubation time.

This book was a tricky undertaking in many respects. In writing about the expedition I wanted to use the voice of 1975, as though I knew nothing beyond that summer. This was easy enough in the beginning, as I adhered strictly to my own records, photographs, and writings of members of my party. But the story quickly outgrew the simple narrative, grew beyond the events and discoveries of that summer, fueled in part by what science had revealed since then. Before long I was writing in two voices: that of the field biologist in the woods, first person and present tense, faced with immediate observations and concerns, and that of the more distant researcher with a broader view and the benefit of additional time and reflection.

I quickly found, too, that I needed help filling the voids where my field notes and those of my colleagues left blank spaces. When I wanted to know, for example, the story behind the airplane on skis that we found in the deep woods, I had to locate someone who could recount the event twenty-five years later. So necessity sent me back on the road, traveling the asphalt and cyber highways, seeking out people that I hadn't been in touch with for a quarter century or never knew in the first place. I found Dave Evans in Arizona, and later, Brad Snow in Colorado, both of whom were living on the Nation River in 1975. And the two of them told me fascinating stories around that airplane amid the spruce trees. When I wanted to know more about Garrett's life, and death, after the expedition and could not find the now-retired Robert Howe who had sent me the field notes that started all this, I located Garrett's daughter, Lisa Jahn-Clough, who graciously told me of her father's difficult years later in life. And by the time I heard about our pilot Gordon MacDonald's accident, the names I previously associated with him were but distant memories. I was, nonetheless, able to contact Dick Hutchinson, still in Circle, and Frank and Marry Warren, retired from the trading post and now mining gold in Central, Alaska, each of whom filled in a few more details and were able to put me in touch with Gordon's widow. For the contributions of all these individuals I am most grateful; and to Lynne MacDonald, especially, I express again my heartfelt thanks for sharing a difficult story with me.

My indebtedness goes still further. A chance meeting with Peter Lapolla, a retired dentist and helicopter pilot in New Mexico, introduced me, albeit many years late, to the essential mechanics of helicopter flight, and a subsequent review of my manuscript by aeronautical engineer David Swartz confirmed my suspicions as to why our overloaded Alouette II was ill-fated from the start. Help came from other directions, as well. Three additional reviewers at the University of Alaska Fairbanks—Terry Chapin, Link Olson, and William Schneider, kept my science and history honest.

Through all of my writing I have benefitted from the influence of other authors, both on and off the page. Most importantly, my brother Philip (one of my favorite writers) has been a constant inspiration and invaluable mentor. Good editors are a writer's best friend, too, and I am especially grateful in this regard to James Engelhardt of the University of Alaska Press. Our many conversations have been interesting, informative, supportive, and fun. And of course, without the original field experience there would have been no story to tell, so to the members of my field expedition I extend my gratitude for great companionship and memories after all these years. Thanks especially to Ed Holsten— you were my strength on the Charley—and to Steven B. Young of the Center for Northern Studies, who gave me the opportunity to explore the upper Yukon and Charley River watersheds in the first place. And for the constant support of my family, then and now, I am especially grateful. You have endured my obsessions so patiently and helped me in more ways than you know.

Just as it takes a village to raise a child (so it is said), it takes much more than an author to write a book. My thanks to everyone, named and unnamed, who has been a part of this endeavor.

LIFE AND TIMES OF A BIG RIVER

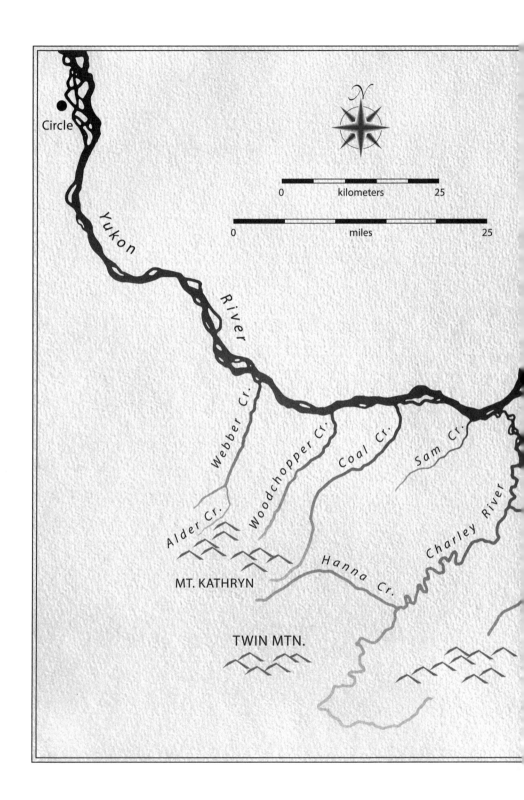

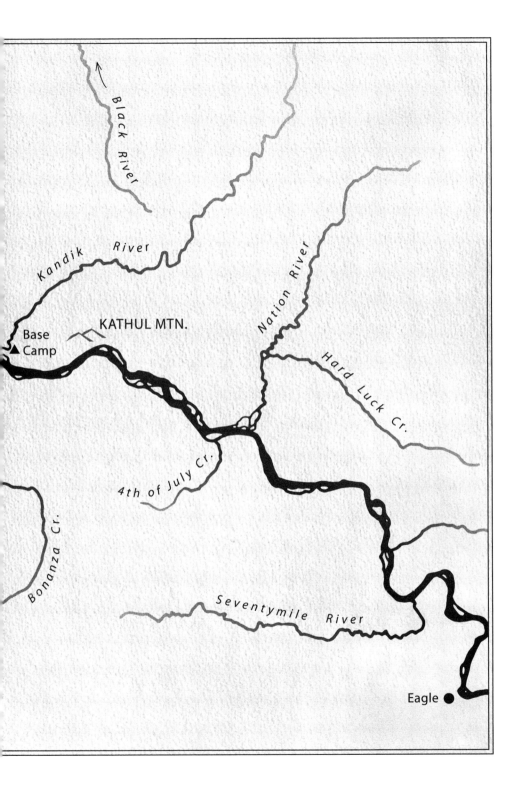

PROLOGUE

In the 1960s and early 1970s, Vietnam was the proving ground for a new generation of helicopter pilots. If your pilot flew in 'Nam, you knew he could handle the aircraft. Mike was a 'Nam graduate, and he was good at the controls. Damn good, I'd have to say. But Alaska isn't 'Nam, and there were two rules of the bush that Mike hadn't fully embraced. Rules that every native bush pilot—men and women who learned to fly growing up the way a farm kid learns to drive a tractor—take seriously. Helicopter or fixed-wing, it doesn't make any difference.

One: don't overload your aircraft. Not when high humidity and turbulence over rough terrain reduces lift capacity to begin with. Not when you're scud-running in bad weather much of the time, flying low through ragged, hanging clouds in conditions that could close in on you at any moment. Flying is tough enough under these circumstances, but finding yourself in a stall because you took off overweight is the stuff of which nightmares are made. Suddenly, the smooth flow of air over the top of your wing breaks up into a turbulent vortex because you're trying to lift too much. You start dropping fast and you can't put your nose down to correct because you're only a few hundred feet off the ground. Low-altitude stalls are almost always fatal. With a helicopter, you're no better off. Increasing the pitch of your rotor blade to maintain lift under heavy load can cause the retreating blade to stall just like an airplane wing, and an overloaded chopper will have less reserve power than a small plane to pull out of it. So experienced bush pilots pay close

1

attention to weight and prioritize the makeup of their load. "Petrol, people, and potatoes, in that order," an old-timer once put it to me. "Simple as that. We'll load as much of your gear as we have room for, after we add up our fuel and body weight."

Two: know where you can get more fuel. Running out of gas when you're forced off your flight plan is a bad reason to end up in the trees. Local bush pilots stash five-gallon cans of avgas all over the place—in the woods by lake shores and remote, brushed-out landing strips; along rivers where they set down on the rocky flats or sandy bars with their oversized balloon tires—wherever they make regular excursions. It's not uncommon to find old, rusting gas cans in the woods lining the big rivers, not a few riddled with bullet holes from days of boredom in the bush, or suspended from wire cables strung around cabins to serve as bear alarms in the night. If you are a local pilot, you know where your emergency reserves are. If you're flying in from elsewhere, you make sure you know which airstrips have fuel. They're on the charts.

Problem is, Mike was neither an old-timer nor a local. Fresh out of 'Nam, his piloting was beyond reproach, but someone else in the army must have done his weight calculations for him. And as for fuel, it's hard to know whether Mike was completely in the dark or just in too much of a hurry to go out of his way for more.

A bunch of guys in the bush on a two-month biological expedition don't travel light. Waiting for air support after our first couple of weeks on the upper Yukon, we had broken camp and stacked all our gear into an impressive pile. Our big canvas wall-tent that would serve as both cook tent and field laboratory was folded flat and formed the base of a pyramid piled high with stoves, cookware, boxes of food, personal gear, research equipment, guns, sleeping bags, and more. We'd been working near the confluence of the Kandik and Yukon Rivers, collecting all manner of information about the forest's past and present, from insect and small mammal populations to tree ring records and ancient fossils. We were ready now to move to higher elevations to continue our field investigations, and helicopter was the only practical way to get there. But Mike apparently was not expecting four of us plus a dog and all our baggage. Then again, I wasn't expecting Mike.

I had made arrangements for this move with a different helicopter service before I left Fairbanks. That was almost three weeks earlier. Once in the bush, communication was reduced to sending messages down the river with passing floaters or taking letters into Circle when we returned there for resupply. Rarely did messages come in the opposite direction, but lack of communication is part of bush life and you learn to deal with it. You learn to deal with delays, changes in plans, silence. You learn patience.

As it turned out, our contracted pilot was having mechanical problems when the time came to move us, so he called another helicopter service, and Mike was sent out. Communication evidently wasn't any better back there, though, or maybe Mike just wasn't paying attention when he got his orders. His surprise now, as he surveyed our gear, was a little disconcerting. He wasn't prepared to break up our load and make two trips into the mountains as I had planned with our original contractor. "Haven't got enough fuel," he said matter-of-factly. We stood there looking at each other while Mike walked around our pile of gear, tapped a couple of boxes with the toe of his boot, as if in so doing he could judge their weight, and thought for a minute adding things up. Then he turned to us with his conclusion: says he was quite sure his machine could make our move in one trip. He said it in the same matter-of-fact voice as before, poker-faced behind his silvered glasses. So, on Mike's word, we lashed the whole pile onto the cargo racks and the four of us, plus the dog, climbed into the cockpit with Mike as if we were weightless.

In a war you take risks. If you survive, you're a hero. If you die, you're still a hero. If you take risks in the bush you may end up in bear feces or rotting in some backwater slough. If you're lucky, someone will find you while you are still identifiable. There's nothing heroic about it. And most people living in the bush are there because they like it and would prefer to stay around for a while. So, they learn quickly what they can get away with—from the mistakes of others, if they're smart—and try to minimize their risks. But flying in the Alaska bush is inherently perilous. There is no accurate weather reporting for remote destinations, and no radar navigation system for vast expanses of the bush, so bush flying is done by visual reference and pilots follow a simple set of guidelines.

Visual flight rules fill a couple of pages in the aviator's manual, but they all boil down to this: never lose sight of the ground.

Never.

Which isn't easy when over the next pass, the ground may be totally shrouded in fog. Flying into clouds without instruments (and few bush aircraft are so equipped) leads quickly to spatial disorientation—a decidedly unhealthy situation.

Mike managed to get us airborne on his second try. Unable to lift vertically, he coaxed the helicopter back to the far end of the river bar, wound the jet engine to the red line, and pushed the cyclic control forward to make a running takeoff ahead of his own blinding dust storm. The willows thrashed wildly as we skimmed over them and veered inland from the river's edge. Beyond the willows, the stiff spires of black spruce poked up at us like the spears of so many ancient and angry warriors. We roared inches beyond their reach, struggling for enough speed to get out in front of our rotor downwash to where the blades could bite into undisturbed air and lift us into the open. Slowly, the muskeg fell away under us.

Mike ran a long way down the right bank of the Yukon before he had gained enough altitude and confidence to try for a crossing. Atmospheric conditions were his concern now. The air was sultry, foreboding. Clouds were building quickly. The stagnant weather pattern that had left us warm and dry for our first two weeks on the river was changing rapidly. The temperature had risen notably the past few days, pushing into the high nineties, and this alone would have the same effect on the aircraft as a substantial increase in altitude, requiring more power and greater lift from our rotor. Higher humidity over the river, almost a mile wide with its braided channels and sloughs and endless waterlogged muskeg, would aggravate the problem. Playing it safe, Mike chose a crossing with a sizable island in the middle of the river.

As we started over the water, the rain came, gently at first, then harder. Once across the Yukon, we needed only to work our way up the Charley River until we found the tributary we wanted and then follow it up to our tree line study site. The Charley was easy enough to navigate, with its broad, dark meanders spreading across the hummocky ochre muskeg, but a dense, gray veil was fast settling over the landscape. Twelve miles upriver we met the canyon of the Charley and its steep sides started closing in. Our simple plan was becoming increasingly difficult with rapidly deteriorating conditions. Convective thunderstorms were charging the air with formidable energy. The strong vertical air currents made for risky low-level flying, and we still had considerable altitude to gain.

Our mountain was now lost in the boiling clouds. The smaller tributaries were blurring through the rain wash on the outside of our plastic bubble, making them difficult to track. Running my finger along the topo map in my lap, we picked a drainage that looked right and started up, leaving the Charley behind. The climb demanded still more of the machine. Mike pulled up on the collective control and rolled in full power. He reduced his forward airspeed, feeling, listening for the telltale shudder that would warn him if his retreating blade was losing lift. We could feel the downdrafts bucking our aircraft. From behind I could see that Eduardo, in the front left seat next to Mike, was tense, alert, glancing at the pilot often for clues. But Mike seemed cool, expressionless, concentrating hard, not talking. Garrett was decidedly nervous; Mark was coming apart, elbows on knees, his head buried in his big hands, eyes closed. Visibility was steadily deteriorating, the rain coming still harder. The confines of the drainage we were following were steepening, narrowing, the rocky creek bed rising beneath us. We weren't gaining altitude fast enough. In the jargon of veteran helicopter pilots, Mike was "all in." He had pushed the aircraft to its limits.

Federal air crash investigators have a language of their own to describe our situation. First, they would say, we were flying VFR into IMC: running by visual flight rules into instrument meteorological conditions, without instruments. And even without the overload, they would note, continuation would lead inescapably to CFIT: controlled flight into terrain. Efficient, emotionless acronyms to describe an event that one rarely walks away from. But Mike had an option not available to small airplane pilots. Boxed in by narrow canyon walls with insufficient room to turn around, he pulled up short of impact, rotors clipping brush on either side of us, and held the aircraft in a tight hover over the creek bed. He yelled at us to jump out. Three of us and the dog bailed, reducing his load instantly.

Standing in the water, looking up into the rain, we watched Mike lift out with Garrett still on board, and in a matter of seconds he was gone.

We waited. An eternity later we heard the helicopter again. Mike reappeared above us, fading in and out of low, fast-moving clouds, hovering, turning downstream then back again like he was looking for something he couldn't find. He disappeared and once more reappeared, higher than before. Then he was gone.

We were alone. We were nowhere. No food, no gear, a lost field party—and somehow we knew we'd never see Mike again. Our mission had been to get to know this land intimately, but now we would get to know ourselves.

I. CIRCLE

July 2, 1975, in a mail plane en route to the village of Circle. Tossing about on the updrafts of a clear day over the Yukon-Tanana uplands, I am intent on the landscape scrolling under me, a complicated mosaic of forest and muskeg, of meandering rivers and cut-off oxbow lakes, of rounded alpine summits and old caribou trails. To my regret, however, I am greatly distracted by a problem I have never encountered before. My pilot is falling asleep.

I could have driven to Circle. The village was connected to Fairbanks and the outside world by a 160-mile, pot-holed gravel road, and the villagers boasted of being the northernmost point in the U.S. that could be reached by car,[1] but it would have been a long day at the wheel after a long trip from the south, and my party was waiting.

Besides, I like to fly. It gives me a much better perspective of my world. I would spend enough time on the ground later, getting to know the land in all its gritty details. So I caught a ride on Air North's mail run and found myself in the company of a pilot who apparently was so bored with his job that it was all he could do to stay awake. That or he'd had a pretty rough night. He was a big man who seemed to slouch naturally—both he and his plane had a very unkempt look—but now he slumped in his seat a little more, crowding the door as if trying to get comfortable for a nap. The two Athabaskan women who had climbed silently into the back of the four-seater don't seem to notice, or at least care much, but I am having a hard time ignoring the situation.

I try conversation above the unmuffled drone of the engine. The pilot isn't making it easy.

"Fly this route much?"

"Unhuh."

"How often?"

"Three times a week."

"Been doin' it long?

"Yup."

"Few years?"

"Four."

On it went, one drawn-out syllable at a time. I was beginning to wonder if it was me. I'd had a conversation like this on my last flight, too, on the plane up from Seattle. Attempting to be sociable, I tried to make conversation with the guy seated next to me.

"What brings you to Alaska?"

"Pipeline."

Then tolerantly he asks, "You?"

"I'm a forest ecologist working for . . ."

It ended right there. Pipeliners and ecologists mix about as well as oil and bay ducks, and the conversation was finished before it had a chance. The man's eyes glazed over and he turned back to the window. I could have explained myself, I suppose—that I was a scientist, not an activist, interested in how natural systems worked—but I knew there was no recovering from that one. I decided to let it go. If the label "ecologist" was going to raise a flag, then I was learning, starting at that very moment, not to call myself one. Not in Alaska anyway.

My pilot was getting visibly irritated by my questions, but the annoyance seemed to be working. Grudgingly, he straightened himself up and lit a cigarette. I let him have his peace and went back to watching the land.

From this vantage point, the Yukon-Tanana uplands bumped on as far as I could see in every direction, low, rounded hills deeply dissected by numerous creeks, the thin forests of spruce fingering their way up the drainages and lower slopes like dark-green dye bleeding into the lighter grey-greens of the matted tundra above. Unspectacular by comparison with the Alaska and Brooks Ranges that define this great interior, the upland summits were little over four thousand feet in elevation. Yet they ascended well above the treeline, where thus bared and devoid of any mark of human habitation, they were given to an impressive desolation, the softened appearance of the alpine tundra deceptively gentle-looking from the air.

The incessant work of freezing and thawing was evident everywhere. In this country where winter temperatures drop to forty or fifty below and often stay there for days, ice becomes the principal element shaping the land, and this through brute force. When water confined in the small pores of soil and rock freezes, it moves mountains, creating a wondrous assortment of landscape patterns not found at temperate latitudes. Not always obvious on the ground, I could see clearly from the air rings of stone arranged through the differential sorting of large and small rock. Countless cycles of frost heaving had lifted the larger stones to the surface of the soil and then nudged them away from centers of frost activity until they joined others to form irregular nets over the broad summits. On the slopes, waves of saturated soil creeping with imperceptible slowness under the mat of vegetation oozed downhill in conspicuous lobes, like heavy frosting on a cake. In the boggy lowlands, large polygonal patterns appeared through the dense cover of moss and shrubs, like the cracks of a drying mud puddle greatly enlarged. Frozen ground, contracting repeatedly under the extreme cold of winter, broke into deep fissures, and every summer melt-water from the surface filtered into these cracks to refreeze, gradually forming thick wedges of ice that defined the polygons.

The undulating hills beneath me were part of a geological formation that stretched from the confluence of the Yukon and Tanana Rivers all the way to the western Yukon Territory. Born from numerous episodes of marine deposition throughout the Paleozoic era, their sedimentary cores were tectonically deformed in the Mesozoic, covered again by nonmarine sediments, and finally subjected to the erosion and Ice Age permutations of more recent geologic times. Under the present cloak of subarctic vegetation, these layered rocks contained one of the richest assemblages of fossil plant and animal remains to be found this side of the British Isles. This was truly one of the greatest libraries in the world; an archive of 600 million years of earth history.

As we skimmed over Twelvemile and Eagle Summits, the land dropped off below us and then submerged gently into an expansive and uninterrupted forest of spruce, stretching forty or more miles toward a dark and level horizon. To the north and quite a ways distant, I caught now and then the white glint of light off water, at first scattered in small reflections, then gradually lengthening into thin strands. In another fifteen minutes or so, the strands themselves began to intertwine in disconnected segments, disappearing and reappearing from a feature-less background. Then, ten more minutes over the last fold in the land-scape, the picture suddenly snapped into focus. The strands connected in their entirety, braiding into a river of startling magnitude.

Straight ahead, the village of Circle materialized out of the spruce muskeg, tucked into a quiet bend of the Yukon away from the tangle of meandering channels and sandy islands. Only a few miles upstream, the great river flows out of its rocky confines to spread out over the country in a confusion of land and water known as the Yukon Flats—sometimes twenty miles or more across—that extends all the way to Fort Yukon and two hundred miles beyond. In this maze even the river seems to lose direction, wandering aimlessly back and forth in endless search of its own course, creating in the process an unsolvable puzzle of the landscape with pieces of constantly changing size and shape. Finding your way in this country can be the ultimate challenge; the first story I would hear in Circle would be of the fisherman from the Lower Forty-Eight who ventured out from the village for an afternoon of angling only to get hopelessly lost, emerging from the bush some-where down the Steese Highway two days later.

My pilot banked a slow arc over the scattered houses at the north end of the village to approach on a light southerly headwind. He set the twin-engine Piper down easily on the gravel runway and taxied up to the oil drums and gas pump at the far end of the strip. In the small group now gathering outside the post, I spotted two of my party.

Nothing less than an act of Congress had brought me to Circle. In 1971, four years before our rendezvous on the banks of the Yukon, legislators in Washington were grappling with a tangle of long-standing aborigi-nal land claims in Alaska, trying to create some semblance of territorial order on a frontier that most lawmakers knew only from the newspa-pers. The roots of this congressional confrontation could be traced all the way back to 1915 and the first organized expression of concern by

Village of Circle, once a boom-town of gold-miners, thought by its founders to have been located right on the Arctic Circle.

Native leaders over the Anglo invasion of their Athabaskan homeland. Ironically, it was two whites, Judge James Wickersham and the Reverend Guy Madara of the Episcopal Church, who called for the meeting with the Athabaskans, ostensibly over concern for the Natives' threatened way of life.[2] The Tanana Chiefs Conference went largely unnoticed then, but it set a precedent, and their ancestral voices would again be heard half a century later.

This time it was the U.S. Army Corps of Engineers who, in 1954, stirred things up with a proposal to dam the Yukon where it narrowed through the Ramparts below Stevens Village. The scheme would have inundated nearly 11,000 square miles of boreal forest, essentially the entire Yukon Flats area, home and subsistence grounds for Athabaskans of seven villages.[3] So the Native leaders organized again and this time found themselves with many allies. The ensuing debate quickly settled on the question of who owned the undeeded land and who had the right to make decisions about it. The Athabaskans finally gained recognition and the dam was defeated, but then oil was discovered at Prudhoe Bay and the stakes changed. With eyes on a trans-Alaska pipeline to quench an insatiable thirst for petroleum in the Lower Forty-Eight, the whole nation became interested in the land ownership issue.

Thanks in large part to their earlier experience with the Corps of Engineers, the Natives were well organized for their next fight. By this time, they were getting plenty of help from their northern neighbors, too. Public outcry over an Atomic Energy Commission proposal to excavate a deep harbor at Cape Thompson, home to the Point Hope Eskimo (through means of an experimental atomic explosion, its potential impact on the Natives dismissed as causing no more disturbance than the excavation of thirty million cubic yards by conventional methods),[4] and the arrest of two Iñupiat hunters by the U.S. Fish and Wildlife Service for putting eider duck on the dinner table (in violation of a new Migratory Bird Treaty with Canada and Mexico),[5] focused considerable attention on the land issues in question. With far more in the offing than preservation of cultural tradition, however, just about everybody got into the act, including conservationists who argued passionately for the protection of some of the last unspoiled land on the continent. The political bargaining finally ended in 1971 when Congress wrote into law the Alaska Native Claims Settlement Act, which decided, at last, the jurisdiction of much of this wild land.

In trying to appease everyone, Congress found itself in a bind, not fully aware of what it owned and not wanting to give away anything it might later regret. So it satisfied the conservationists for the time being by writing in a provision that one hundred million acres or so of designated lands in Alaska should be studied for possible inclusion into the national preserve system. One of those designated lands was the area drained by the Charley River and neighboring tributaries of the Yukon: a vast tract of interior Alaska boreal forest and tundra, over four thousand square miles, inhabited by perhaps no more than two dozen homesteaders—though no one knew for sure—and without a single road penetrating it. I was part of the study team, headed into that empty map for two months to conduct the first systematic biological inventory of a forgotten place.

Garrett and Bruce met me with robust handshakes. "Glad you made it! We weren't sure whether you'd be on the plane or not."

"Great to see you guys. Yeah, I got lucky. I went straight over to Air North as soon as I landed in Fairbanks and reserved the last seat. I hoped I wouldn't have to spend more than one night in town."

"Good flight up? We heard there was some weather coming in after we left Seattle."

"Smooth all the way. Are Mark and Eduardo here?"

"Yeah, they're walking around somewhere. Should be back any minute. They had a bit of a problem a couple days ago, but they'll tell you about it."

We chatted energetically, bouncing brief quips back and forth, watching the pilot wrestle his cargo out of the rear compartment and snatching up my boxes and duffels as he dropped them on the ground. Having come in on the previous mail flight, Garrett and Bruce were awaiting only my arrival, and having quickly exhausted the sightseeing possibilities in Circle, were anxious to get on the river. We each grabbed an armload and shuffled over to the trading post.

Circle was a quiet village of about seventy residents, fifteen or so Anglo and the rest Athabaskan, hanging on primarily as an outpost for river travelers and a handful of homesteaders along the Yukon. The trading post was not the weathered, hand-hewn log building I had imagined, but rather sported the prefabricated look of the 1960s. The simple facade of unnaturally uniform logs was distinguished only by two unpainted aluminum doors at opposite corners of the storefront. A sign over the left one identified it as the "Yukon Liquor Cache," and over the right, the "U.S. Post Office, Circle City, Alaska." Both doors opened into the same room. The smaller print on the outside advertised groceries, gas and oil, souvenirs, raw furs, ice, and tire repair: the basic necessities here. But the sterility of the building's manufactured look and uncluttered business facade belied the interior warmth and social importance of this little outpost in the bush. Frank and Mary Warren, the owners, provided just about every service that was needed in the country, from hot meals and showers to emergency radio and air taxi service. Inside, the post was crammed full of essentials: flour and sugar, oarlocks and ammunition, dishpans, gold pans, frying pans, fish lures, dip nets, and black rubber boots (size large). But canned goods were the specialty—canned milk, meat, fruit, fish, crackers, jam, pea-nut butter—goods with indeterminate shelf life, destined for long stor-age in isolated cabins and high caches scattered throughout the bush; dependable goods to stave the hunger of frigid nights on the trapline, if the bears didn't get them first.

Here was a lesson in elementary provisioning—just basic goods at land's end prices. But it was also a reflection of a way of life that was decidedly Spartan. If not far from the influence of mainstream Amer-ica, it was at least far from the marketplace, and it showed not just on the shelves but in a general air of resourcefulness that pervaded the post. When I asked matter-of-factly if the post had a pencil sharpener that I might use, a gruff clerk cast a stern eye at me, tongue-tied. I could tell he was searching for choice words, but he spared me and instead politely asked if I weren't carrying a pocket knife. Of course I was, and was immediately embarrassed by my question. And I knew I had been let off easily by this woodsman. I thought about this often in the days ahead; it was a simple incident, but an important one. That little jolt did much to arouse my sensitivities to the cultural place.

Outside, the day was bright and the air warming rapidly. Bruce, Gar-rett, and I loitered in the sun with coffee cups steaming, waiting for Mark and Eduardo to show up. Our conversation had mellowed con-siderably from the excited pitch of our first meeting, turning mostly to logistical queries of one or another as we scuffled a little nervously in the gravel.

"Mark say anything about the base camp?"

"Guess they found a good spot. They got everything set up."

"Have you checked out the boat?"

"Yeah. Looks okay. Not the prettiest boat on the river, but it floats."

"Enough room for all this stuff?"

"Hell, you oughta see the pile of gear I've got."

Our mood was decidedly less festive now as each of us pulled back into our own thoughts, running through mental checklists for the ump-teenth time in anticipation of our momentary departure. We had pre-pared ourselves as best we could for this day, yet as we stood now on the rough-cut edge of the Alaska bush, many anxieties welled to the sur-face. Our mission was to collect as much information about the natural history of the upper Yukon and Charley River area as we could, but our collective experience was skewed heavily toward the academic, and we knew we would be tested in the coming weeks.

With divided attention, I surveyed the village scene around me, curious about the lifestyle of the locals while also hoping to spot the other two. It was a quaint place, with all of the feeling I had expected of a village on the edge of contemporary society. Most of its original log buildings are gone now. As Hudson Stuck, explorer-missionary for the Episcopal Church, observed a half-century earlier after the last wave of miners had cleared out, "in this country of extreme cold and roaring stoves, there is always a useful way to dispose of deserted cabins."[6] Today's houses were small, mostly wood frame, and weathered to an appearance much like that of any coastal fishing village battered by time and the elements. Every yard was a museum of artifacts, an odd assortment of old and not-quite-new: buck saws and chainsaws, broken sleds and Honda three-wheelers, a Johnson outboard mounted on an old oil drum, burned-out wood stoves and assorted stove pipes, a Sears wringer washer with clothes on the line.

A few sled dogs, thin and mangy-looking, neglected in their off season, lay chained about, dreaming of the return of winter when they would once again command respect and attention. But what these dogs didn't understand was that they were living in the past, kept on now by only a few mushers in the village who still relished the sport—if not the utility—of dog-sledding. The fate of the dogs was cast in the iron and fiberglass hulks that also waited nearby, half hidden under the clutter of summertime. With the new gasoline-powered snowmobiles a man could tend a trapline of three or four times the length covered by his dog team, though it was not without some sacrifice. Rarely could a broken part on the motorized sled be repaired in the bush with sinew, or babiche, or birch splints. So the trade-off was a degree of independence, and not an inconsequential one at that. Most of the Natives, though, had come too far already in relinquishing former skills for the new technology, and now they were caught in that uncomfortable limbo somewhere between their ancestors' world and their children's.

All the buildings in the village, save the new trading post, showed the wear of a land that strained endlessly to reconcile the extremes of a northern continental climate. Under the relentless cold of winter, the earth buckled and heaved with the expansion of freezing water, the weight of a house no measure of counter force, and in the summer heat the land settled back into a quagmire of saturated soil. Permafrost—the perennially frozen ground lying immovably and impenetrably only a few feet beneath the surface—complicated matters immeasurably. Heat bouncing off the sunlit wall of a building or conducted downward from its foundation would thaw the frozen soil more on one side than the other to pitch the house deeper into the ground toward the south, giving the building the appearance of having been raised by a well-oiled construction gang. Beams twisted, roofs sagged, and pencils rolled off the kitchen table, though the occupants seemed not to notice anymore, or at least not to care much. The eastern Siberians who had coped with this sort of thing long before Circle City was settled built houses on flexible foundations to absorb the stress. They also left their basements open all winter to encourage deep freezing, while closing them up tightly during summer to prevent thawing. That which remained frozen permanently did not move.

If the climate were showing its effects on man and machine, so, too, was a shortage of materials for maintenance. Everywhere I looked, some mechanical contraption, fully exposed to the weather, was lying in wait for a knowledgeable hand with the right-sized bolt, belt, water pump, or whatever. I guessed that probably half of the machinery in the village was currently being cannibalized for parts to keep the other half going. So the village had, by Lower Forty-Eight standards, a poor and trashy look about it. And yet there was a certain atmosphere of self-determination that appealed to my senses as I looked around. Like so many other villages in the far North, remote and isolated for a good part of the year, Circle existed only for itself, striving for nothing more than rudimentary comforts. It did not have to please anyone but its own citizens.

At the far end of the only road, two stocky men were making their way slowly toward us, working the empty barrels and vacant back yards like L.A. street people. I came to my senses when I recognized one of them as Mark, my field assistant. The other I presumed then was Eduardo, whom I hadn't yet met. As soon as Mark spotted me, he and Eduardo abandoned whatever it was they were up to and picked up the pace. I walked quickly toward them.

"Hey Mark, finally caught up with you. Eduardo? Hi. I'm Peter. Been looking forward to meeting you. Heard lots about you . . . not all bad, either."

He laughed. "Well, you don't know me yet. Nice to meet you."

Eduardo was up from Seattle. He was the only one among us who was familiar with the country, having worked the river himself the year before, and was the one I would be looking to most for leadership in the field. Mark was an Easterner like the rest of us and had joined Eduardo in Seattle, at my recommendation, to accompany him on the long drive to Fairbanks with our equipment. Things hadn't worked out quite as I had hoped, however, for they were not hitting it off well. Their three weeks together had strained what was by now a touchy relationship, with a long field season yet to go. Nevertheless, keeping peace as best they could, the two of them made it to Fairbanks loaded down in Ed's old Chevy truck, picked up a boat there and hauled it the six-and-a-half hour drive over the Steese Highway to Circle, and had already been upriver to establish our base camp. Three nights ago they came back down to Circle to pick up more gas, groceries, and the rest of us.

"How's everything going? Garrett tells me you got a base camp all set up. On the Kandik?"

"Yeah, seems a good spot. I was pretty excited about it, lookin' forward to a good summer. . . 'til a couple days ago."

"So what happened to change things?"

"Ah, just some bastards here ripped us off and nobody seems to know anything. Took our tent, sleeping pads, and fishing gear the minute we turned our backs on the boat. We've had to sleep in Frank's old storage shed—his bunkroom's full right now—and the mosquitoes are horrendous."

Mark stood quietly, letting Eduardo do the talking.

"The longer I listened to the goddamned mosquitoes that first night, the madder I got, so in the morning we went over to this guy Albert's house to see if he could help. Heh! Foolish me!"

"Albert?"

"He's the village leader, so I'm told. Not a very big guy, but feisty as all hell you know. I tried to be diplomatic, merely suggesting that because he had many contacts maybe he could help us recover our gear. But he gets all huffy, asking if we were insinuating that it was his people that stole the stuff. I tried again tactfully, but he just reached over for his .30-06 standing in the corner—he was sittin' at his kitchen table—and says we're not welcome in his house."

"Pisses me off," he added. "You know damn well whoever ripped us off doesn't need the stuff. They just don't want us here. Rumors are already goin' around that the government wants to establish some kind of a preserve here and everybody's getting uptight. Gonna take it out on us, I guess. So we've been poking around in backyards while we were waiting for you, hoping to find the gear stuffed in a barrel somewhere."

Garrett and Bruce were standing beside me, listening to the story again.

"Can we get along without that gear?"

"I guess. We've got some back-up."

"Well then, why don't we just get on the river. Any need to stay longer?"

Eduardo led us down to the bank where the boat was tied. A lean, bespectacled gold panner from Texas was keeping an eye on things for us, determined himself to stay all morning on the gravelly shore if that's what it took to sift a fleck or two of gold out of the murky water. He noted with a wry drawl that he didn't much care whether he got rich, just wanted to say he'd done it. Eduardo showed him again how to swirl the pan. Two young Athabaskan boys, barely tall enough to reach the Texan's pockets, had annoyed him to no end with questions about the contents of our boat but then turned shy the moment we arrived. I'd have passed their curiosity off as typical of children, but Ed muttered something under his breath about the cute little devils probably having big brothers who'd give 'em a nickel for their information. The boys watched us intently, dark eyes and beautiful brown skin, black hair shining in the sun.

The twenty-foot, flat-bottomed, aluminum barge in front of us would be our supply link to Circle for the next two months. Definitely utilitarian, the boat showed more than enough wear to have demonstrated its worthiness on the river. Still, I studied it the way I supposed the Athabaskans eyed the first sternwheeler to paddle its way up the Yukon, with a kind of native suspicion or what I prefer to think of as scientific objectivity. Our boat was powered by a forty-horse outboard

Navigating the Yukon with a 55-gallon gas tank and spare motor. Photo of the author by Ed Holsten.

motor, mounted on a mechanical lift so the propeller could be brought vertically upward and nearly out of the water to navigate the shallowest of channels. A fuel line ran directly to a fifty-five-gallon drum lying on its side for a gas tank. A spare fifteen-horse outboard sat on the bottom of the boat, lashed to D-rings on the transom.

With its square end and extra-wide beam, the boat could carry a payload of nearly two thousand pounds, most of which at the moment was extra gasoline. Eduardo and Mark had taken on fuel in every container they could find, and everything else that needed to go upriver had to find a place on, around, or between the gas cans. A bit disconcerted now by the amount of personal belongings I was traveling with, I worked hard to tuck my gear in wherever I could find space. And as I labored at it, I also snooped in the grocery boxes, for I too was curious about the contents. Not bad, I thought. Looks like we'll get by alright for a while. And what's this here? Egad, a whole case of Cutter insect repellent in spray cans.

II. FLAT WATER

FIVE MEN AND A DOG ✦ DANCING FOREST ✦
KUTCHIN UNDER SEIGE ✦ SALMON FOR THE SWEDES

We pushed off that afternoon into a future no one could have predicted, five men of hardly more than accidental acquaintance whose only stake in this venture was some peculiar brand of scientific interest. We were an eclectic group, as much a mix in appearance and personalities as in expertise. Garrett was a mammalogist, and if any of us looked the part of a field scientist, it was he. His thick, wavy hair and heavy, reddish-brown beard beneath prominent cheekbones hid a face that was thin and honest in its features. He was a slight man, fit and wiry, with a ready smile and constant twinkle in his eye, and he carried an air about him of comfortable ease in the wild, of true *gens du bois*. I had known Garrett from a brief collaboration the previous winter in northern Vermont, but it was enough to see that he was a competent field biologist. Garrett's work had already taken him far afield, from lemmings in Norway to hutias in the Caribbean, and he seemed to have a sixth sense about the animals he studied. I was sure that nothing would escape his notice on this trip.

Bruce, more than any of the others, upheld the academic image of the group. He came from a family of learned men and had accepted the role with a certain gusto and self-satisfaction. Young and just getting started on his doctorate, he was a brilliant student of paleontology, already making waves in the Ivy League. Bruce was tall and lanky, all arms and legs—an Abe Lincoln sort, given a studious look by his glasses. In amusing contradiction, he wore a dark green commando hat with the brim on one side flipped straight up and held by the braided neck-cord that extended over the top. Bruce was a thinker, and whenever he spoke his words were carefully measured, clearly enunciated, and delivered from the back of his throat with an air of authority that belied his young face. A friend of a friend had recommended him highly, and though this was his first expedition of any sort, I could see that he had done his homework and took his role seriously.

Mark was the youngest of the group: a big, muscular lad, blonde and fair skinned, whose physical strength I thought would be an asset to the expedition. He was an undergraduate where I had done my doctoral work back East, and had accompanied me on several winter excursions up Mount Washington during some of my earlier research. His mountaineering skills were well practiced, and in addition he had been through considerable outdoor leadership training. This seemed better than usual preparation for the kind of expedition we were mounting, so I hired him on as my field assistant. However, I was now concerned about his apparent lack of enthusiasm as we prepared to start upriver. In the company of the rest, whose excitement was contagious, Mark's quietness stood out; but then he had already been working for three weeks and undoubtedly some of the luster of the expedition had worn off. I was troubled, too, about the tension between he and Eduardo.

Eduardo had a few years on Mark and though he never demanded or expected it, his experience in the bush commanded a certain respect. Ed was a rugged sort, but with a naturally calm and purposeful manner about him that was reassuring, and he quickly won the confidence of the others. Eduardo wasn't his real name, but because he had spent a few years in the Peace Corps in Chile, followed by a research stint in Costa Rica, and because he had the dark eyes, hair, and complexion of a Latino, the name had fit well and stuck. Whenever he was addressed by it, he would grin and flash beautiful white teeth beneath a coal-black moustache. Eduardo was never without his scruffy brown felt hat with the squirrel-tail band, and in the bush he always carried a double-barreled shotgun, its stock thrust back over his shoulder and his fist around the muzzle, giving him a rather rough look. His dog, a young black Labrador retriever named "Moose," was seldom far from his side.

One would not have guessed from outward appearances that he, too, was a man with academic credentials, a doctoral candidate in entomology at Washington State University. Eduardo was unpretentious and comfortable anywhere, traits that I suppose had a lot to do with his Peace Corps experience, and I was glad for his presence.

This was an opportunity I had aspired to for a long time. I had always been drawn to northern places and had just spent the better part of my twenties at the University of New Hampshire, getting serious about boreal forests, bogs, and tundra. Fresh out of school now, I'd had an odd mix of training, starting out in earth science and hydrology, but ending up with an advanced degree in botany and plant ecology. Not comfortable with large organizations, I fell into my first job as a result of a brief conversation following an afternoon lecture by a Harvard-trained taxonomist, one who was equally enamored with the North. Steven Young was starting his own Center for Northern Studies in the highlands of the Vermont-Canadian borderland and asked me to join him. Without considering another opportunity, I picked up my family and moved. Steve had already laid the groundwork for the present expedition, having won a contract with the Department of Interior a year earlier, and within months I was on the Yukon.

Eduardo untied the boat and eased it off the gravelly shore, into the shallows. It bumped and rocked awkwardly as Bruce, Garrett, and Mark shifted their weight in an effort to help the stern off the river bottom. The dog stood with classic Lab conformation, ears perked and front feet up on the transom, never more than half-a-whistle from jumping overboard. Eduardo and I, on opposite sides of the boat, waded into the water until we were knee deep, and gave a hard, protracted shove into the current. As the inertia of the heavy boat jerked me into deeper water, I lunged ungracefully for the gunwale, groping and sliding feet up and face down into the boat. We were underway.

Navigating our way out of Circle was a task that required all of our attention, for the reputation of the Flats is well-deserved, and the main channel is not gained easily from the village. Drawing more than usual with our full load, we felt our way carefully along the braided waterway, reading the ripples and swirls of the surface for their clues to the unseen bottom. Eduardo skillfully picked his way through the shallows, idling little faster than the opposing current, at times lifting the prop nearly out of the water to skim across the shifting sands. But I was more intent on our wake.

I watched the last house of the village flicker in the trees and disappear, like a dying flame sucked into the darkening forest. I tried hard to keep it in sight longer, but it slipped away from me as easily as a dream upon awakening; and then Circle, the end of the communication line for us, was gone. I clutched for every detail of the shoreline behind us now, knowing that I would one day have to return on my own to find Circle in this maze; mindful that out in the main channel, a person could drift right past the village without a hint, without hearing the dogs or catching the scent of wood smoke, until an hour or more downstream when the village seemed long overdue. "Them blamed Flats is the meanest part of the river," an old pole-boater once opined. ". . . A man don't know where to go."[1]

When we finally turned into the open expanse of water, I let go of Circle and relaxed, leaving navigation to the others, and settled into the marginally comfortable contours of gas cans and duffle bags. The water was a cold, steely, silver-brown in the late afternoon light, thick with silt and as opaque as the ground it flowed over. The principal source of the upper Yukon's sediment is the White River, which tumbles out of the glacial valleys on the eastern flanks of the Wrangell Mountains, several hundred miles to the south. By a circuitous trick of hydrography, the White carries this Alaskan soil from its near-coastal origin eastward into Canada, then north for a couple hundred miles and eastward again, until it discharges into the Yukon a hundred or so miles above Dawson. Thence the soil is soon returned to Alaska to be deposited, shifted about, and picked up again as the Yukon flows westward the entire width of the state. Mesmerized by the sparkle of light off water, I delighted for a while in this quirk by which the southern slopes of the Wrangell Mountains fall directly into the Gulf of Alaska, but soil from its eastern slopes travels westward two thousand miles, all the way to the Bering Sea.

I was aware, however, that my intellectual delight was not shared by others. In some eyes, the Yukon has been forever ruined by the White. "I never saw an uglier river," wrote Alexander Murray when he came upon the Yukon in 1847.[2] Hudson Stuck did not mince words either, as he admitted to having a "foolish resentment of the White River:"

> Glaciers must be drained, no doubt, but one wishes they could find some subterranean sewer to the sea; or, since the Tanana is already their special conduit, and no conceivable addition to its waters could make them fouler—the drainage of London would not affect their colour, consistency, or potability—one regrets that some slight change in the elevation of land did not discharge both sides of the glacier-bearing mountains into the same channel.[3]

Murray and Stuck were entitled to their opinions, but I found it difficult to condemn the river for the burden it carried. What grander purpose has a river than to ferry mountains to the sea? The low sun now shone upward into the clouds, reflecting pure gold onto the water.

The Yukon in all its meanderings is more than a mile wide upriver from Circle, but soon narrows between steeper banks, bringing us into more intimate contact with the land. The riverbanks were sheer, in places deeply undermined by the flowing water. The soil just a half-meter below the ground surface was still frozen, even in the warmth of July. It had been so for much of the last millennium and maybe a lot longer. The surface layer would likely thaw a little deeper yet, perhaps down to a meter, but then summer would wane and everything would freeze up again. As the river continued to erode its bank, the perennially frozen soil would resist collapse, giving to undercutting until the thick mat of vegetation above eventually slumped over the edge. With many of their roots still firmly anchored, the black cottonwood and tall, straight white spruce along the bank would gradually lean out over the river until they stretched horizontally to comb the flowing water. Lieutenant Frederick Schwatka of the U.S. Army, described these leaners as "a series of chevaux de fries or abatis [the latter, according to my dictionary, 'an obstacle of trees with bent or sharpened branches directed toward the enemy'] to which is given the backwoods cognomen of sweepers, and a man on the upper side of a raft plunging through them in a swift current almost wishes himself a beaver or a muskrat so that he can dive out and escape."[4] He wasn't overstating the case by much.

Calving of frozen riverbanks undercut by the Yukon current.

These sweepers could pluck a person right out of their boat if they were careless enough to get caught in the current too close to shore. Eventually, the overhanging banks would get too heavy for themselves and massive blocks of shoreline would collapse into the water with a thunderous boom, like calving icebergs.

Further from the river's edge, the permafrost lay closer to the surface, so that the roots of white spruce could no longer find sufficient growing room in the shallow-thawed soil. Here, the forest opened out into boggy, moss- and shrub-dominated muskeg; spongy and wet underfoot, and a mosquito haven. And it is here that black spruce made its stand, where it was uncrowded by others less tolerant of the cold, waterlogged, and nutrient deficient soils. As trees go, black spruce is a bit of a poor excuse, stunted in height and with little more than thickly foliated twigs for lateral branches. But it is a survivor nonetheless, hanging on

Ice-wedge (see p. 9) exposed with the collapse of the riverbank

where other species are incapable, right to the very climatic limits of tree growth. Scrawny though they were, each tree in this muskeg cast just enough shade on the cold ground beneath it to allow the formation of a lens of ice under its roots. In time the lenses grew into mounds that tossed the trees in one direction or another, at odd angles, imparting a chaotic disorder to the stand. The drunken forest it is called. But if you are of a more positive frame of mind, it is the dancing forest, the trees draped with hanging lichens swaying to an ancient score, choreographed in a punctuated, time-lapse rhythm. The music ends when the mature tree finally topples from its mound. Sunlight then floods the ground, driving the ice back, and a new seedling establishes itself to begin the next dance in an endless cycle.

From what I could see of the forest edge it seemed a poorer land than the Flats, yet I knew that in the long days of the subarctic summer, enough energy flowed through the forest community to support an abundance of life year round. In a few weeks' time, the open muskeg would be fairly bursting with fruit. The thin, creeping runners of cranberry that thread their way through the feathery mosses would hold forth large wine-colored propagules, which would turn juicy and sweeten with the first freeze of autumn. Dense thickets of blueberry would produce enough fruit that bears would come down out of the high country to rake them into ever-wanting mouths. The sedges of the wet ground would yield leaf stalks, seeds, and rhizomes sufficient to feed many voles, and the voles would keep the foxes and weasels from going hungry. The willows, in a flurry of summer growth, would convert enough sunlight to cellulose to support the huge moose, and one fit moose will feed many wolves, or a human for a good part of the winter.

This land once belonged to the Kutcha Kutchin, an Athabaskan people whose traditional subsistence grounds extended from Tacoma Bluff, upriver, where it met the neighboring Han territory, to the confluence of Birch Creek and the Yukon, a hundred miles downstream.[5] Here, generation after generation, their lives moved in rhythm with the shifting resources of the boreal forest. Using post-and-withe weirs and woven basket traps, they tapped the summer spawning tides until their drying racks sagged under the weight of split salmon. When the last King had made its autumn run, the fish camps quieted and the people looked back toward the land, where berries were ripening in the bogs and the animals were beginning to show fall restlessness. The men and boys picked up their longbows and pursued waterfowl with blunt arrows on the stillwater sloughs.[6] With swift and silent canoes, in the crispening fall air, they hunted the backwaters of the river for moose and caribou. And when the early snowcover and river ice was still too thin for efficient overland travel, they rested, and prepared for a winter of trapping—not for a European fur market they did not know, but for themselves, for food and warm covers.

It was a hard land in many respects, as hard a land as any indigenous northern peoples ever endured. Winter here is a time of staggering energy deficit, the searing cold of the dark heavens above like a vacuum that sucks the heat out of every animate molecule, drawing the very life out of breath and leaving it hanging in the air in frozen suspension. At sixty-five below zero, living trees explode under the strain of the cold, and the snow itself, dry and brittle, squeals in complaint underfoot. But all this the Kutchin accepted. If the land were locked in deep-frozen isolation, they learned to make sleds and intricate snowshoes, and traveled farther cross-country than was ever possible over the sodden summer ground.[7] And if the long winter darkness confined them to their lodges, they devoted time to each other, strengthening family ties and perpetuating oral traditions through long hours of visiting and storytelling.

If it were a hard land, it was also a benevolent land, for it amply supplied the material needs of these aboriginal people. The expansive wetlands of the Yukon created a habitat for a rich assemblage of plants and animals—rich, at least, for this part of the country. And the lifeway of the Kutchin in this land embodied not only the physical skills necessary to harvest this energy, but a spiritual ideology equally necessary to balance their relationship to the land. To these people, the physical and spiritual realms were inseparable. Every element of the landscape, every natural entity, had a spirit of its own, had eyes, and watched.[8] And so, guided by a strict moral code, the Kutchin respected the land, appeased the spirits, tabooed waste, and passed with scarcely a mark; without monuments, without landfills, without scars. The material traces of their past are largely gone now—gone with the freshets of spring runoff and the shifting sands of the great river; melted into the soft and silent mosses of the forest floor. The remaining Kutchin have been largely absorbed into the white culture that succeeded them, but their past is not forgotten.[9]

What we know firsthand about these quiet Athabaskan people, before they were acculturated into Anglo ways, we know only from the writings of a handful of French and English explorers who drifted the wild waterways, seeking to expand the fur trade for the Hudson's Bay Company. Among them was Alexander Murray, who established Fort Yukon as a trading post in 1847, trespassing on Russian territory ("We are over the edge," he admitted, "and that by a long chalk, which I call six degrees of longitude across the Russian boundary."[10]); a young English adventurer, writer, and artist named Frederick Whymper, who signed on with the Western Union company in 1865 to survey a telegraph route across Russian America (this time with permission of the Russian government); and later Schwatka, who, in 1883, was sent to gather information on the Indians of the Yukon River for military interests (an illegal mission according to Schwatka, kept as secret as possible to avoid being recalled. The expedition, he wrote, "stole away like a thief in the night and with far less money in its hands to conduct it through its long journey than was afterward appropriated by Congress to publish its report."[11])

To a person, these early explorers described the natives of the upper Yukon as gentle and friendly. In Whymper's words, they were "a harmless, inoffensive set of Indians, cleaner and better mannered than the 'Russian' Indians," and "ever ready and willing to render every assistance they can to whites."[12] But the irreversible process of change for the aborigines was well underway even then, as the English traders found evidence among the Kutchin of earlier Anglo contact: artifacts from Russian trade with the Athabaskans of the Cook Inlet and Lake Iliamna area, and the Russian trading post at Nulato, had filtered this far inland. The indelible stamp of progress had already fallen with a silent thud. When the Reverend Robert McDonald of the Church of England arrived at Fort Yukon in 1862, with holy visions and a mind to translate the Gospel into the native language, acculturation took on a deliberate purpose.

Then came the gold seekers. By 1873, the first prospectors to see this country had probed their way in through the Mackenzie and Porcupine drainages, the latter to its confluence with the Yukon. Thirteen years later, a major strike on the Fortymile tributary brought hoards more into the country. In progressive waves, they encroached upon the aborigines: 1888, gold on the Seventymile; 1892, the Mastodon Creek strike; 1893, Birch Creek. The latter was the richest yet, and when word hit the Fortymile, miners there dropped everything, packed up what they could carry, and scrambled for a bigger dream.

At the closest point on the Yukon to the Birch Creek diggings, a camp was set up. It might have become just another squatter's shanty town on the frozen bank of the Yukon had it not been for the attention of Leroy McQuesten, who might properly be called the father of Circle City. McQuesten was a born trader who spent his boyhood on the Upper Peninsula of Michigan, and earned his spurs as a young voyageur for the Hudson's Bay Company.[13] Eventually coming into the employ of the Alaska Commercial Company, he picked up a few business skills and then struck out on his own, following the Yukon gold-boomers wherever they needed a supply post, and wherever he could turn a profit. His intentions were honorable though, and the man was well liked throughout the region. It was he who, along with Alfred Mayo and Arthur Harper, had earlier founded the trading post known as Fort Reliance, below Dawson, and it was there that he built his reputation. The trio were described as "[typical] frontiersmen, absolutely honest, without a semblance of fear of anything, and to a great extent childlike in their implicit faith in human nature, looking on their fellow pioneers as being equally honest as themselves."[14] When gold was discovered on the Stewart River, McQuesten followed; thence on to the Fortymile (named for being forty miles below his Fort Reliance post). And it was he who grubstaked Sarosky and Pitka, the two miners who discovered gold on the Mastodon and opened up the Birch Creek area.[15] So naturally, when the miners jumped to the next promise, McQuesten and company moved down from the Fortymile. A two-story log building was raised on the high bank of the Yukon, and Circle City was born.

What followed was a quiet massacre. Prospectors crawled all over the Kutchin homeland. There was no bloodshed; the Anglos simply took over, empowered by a foreign law that allowed them to lay legal claim to whatever they wanted. That the land was already occupied by another people seemed not to matter to anyone. The miners had the backing of the U.S. government, and that was the only reality.

Powerless, the Kutchin could do little but watch their own cultural demise. After several millennia of stewardship, they and their close neighbors, the Birch Creek Kutchin and Han, found that they had no rights to the land.

By 1896, Circle City had grown to a settlement of three thousand people, replete with twenty-seven saloons and eight dance halls. Four hundred log buildings lined its main street in what was said to be the largest log cabin city in the world. The miners were followed by preachers and prostitutes, schoolteachers and businessmen, many finding it easier to mine gold from the sourdoughs than from hard rock and frozen ground. A library was built, a theater, a school, and Circle City was looking and acting like any boomtown in the Lower Forty-Eight.[16] Natives attracted by the trappings of an alien society built homes and hopes on the periphery. A few tried to ride the tailwind of the frenzied gold rush, but the future was moving too fast for them and they were no match for the Anglo businessmen. Then came news of the Klondike strike and almost as quickly as the wave had hit, Circle emptied out and the land fell quiet again, leaving the Natives a little poorer. Hudson Stuck, who was acutely aware of even his own influence on Native culture, summed up the situation with characteristic sensitivity: "The tide of white men that has flowed into [the] Indian neighborhood gradually ebbs away and leaves the Indian behind with new habits, with new desires, with new diseases, with new vices, and with a varied assortment of illegitimate half-breed children to support. The Indian remains, usually in diminished numbers, with impaired character, with lowered physique, with the tag ends of the white man's blackguardism as his chief acquirement in English—but he remains."[17]

It was well into the evening and getting quite chilly. Everyone but Eduardo couched lower into the crevices of our cargo to doze. I pulled my collar higher and retreated deeper into my jacket pockets, but for fear of missing something, I couldn't have slept if I'd been hit over the head. In my first few hours on the river, I had already seen three moose, and only a couple of miles back we had surprised a grizzly at close range. The bear was swimming the channel and just about halfway across when we popped around the corner. We both reacted with equal urgency. Eduardo throttled back immediately, but we were carrying a lot of forward momentum and it seemed as if the river current was going to sweep that bear right into our boat. For its part, the terrified bear never gave a thought to continuing its course. It clawed frantically at the water and mustered a surprising amount of speed toward the shore it had just left, where it hit the river bank at a full gallop and never looked back. That was about as close as I ever wanted to get to a grizzly.

I had already stopped counting bald eagles, though I never tired of seeing another. Earlier, I watched one swoop down over the water to pick up a large floating stick, and with my binoculars I followed it back to its nest: an enormous platform, several feet in thickness, perhaps seven feet in width, the older layers grown over with lichens and moss. Minutes later, its mate took off along the shoreline, gliding through the trees, attempting twice to snatch dead branches on the wing. I wondered how many generations of eagles had remodeled that nest since the first.

As the channel deepened, I spelled Eduardo at the helm. We droned on, transfixed by the endless line of trees on either shore, measuring our progress by the passage of creeks and cabins, river bluffs and river bends. Place names on the map brought to life the trials and successes of earlier immigrants—creek names like Eureka, Thanksgiving, Hard Luck, Easy Moose, Poverty. The cabins, too, spoke of high hopes and hard times, reflecting varying levels of skill but a common determination on the part of their builders. The sod-covered roofs of earlier days had mostly collapsed along with the played out gold rush, but the return of subsistence living by a few Anglos saw a handful of the better cabins refurbished with new tin roofs. They were vacant now as their occupants had moved to fish camps or to town, seeking cash income during the summer months. Weeds grew tall in front of the doorways but would hardly be noticed under the autumn frost when the owners returned to take up the rifle and traps again.

Thirty miles upriver from Circle, the left bank of the Yukon bowed gracefully around a blunt promontory of resistant rock. Just beyond, hidden from view downstream, the procession of spruce on the river's edge was broken explosively by towering volcanic cliffs. Three hundred and eighty million years earlier, more or less, in the middle of the Devonian period, great pinnacles of molten basalt and yellow tuff erupted from the ancient sea floor beneath and shot upward three

High bluffs of the Woodchopper Volcanic Formation, upriver from Circle.

hundred feet to freeze in place. It was a startling change from the monotonous horizon of the Flats: the jagged, sharply eroded skyline suddenly soaring above the spruce, mirroring their spire forms with roughly textured rock towers and deep clefts. For the next twenty miles this rock formation, the Woodchopper Volcanics, would produce prominent bluffs along the river. And within this igneous outburst, enigmatic exposures of sedimentary marine limestone (the middle of the Devonian period was also the middle of the Paleozoic era—the age of fishes) taunted the mind with misplaced remains of corals, brachiopods, and numerous other invertebrate fossils. This will not be a dull summer for field science, I thought smugly.

At the mouth of Coal Creek we stopped to talk with some homesteaders, two couples, that Eduardo and Mark had met three days earlier. Known on the river as "the Swedes," they were actually three Swedes and one American. Dale, the American, had moved to Sweden, met his wife there, and sometime later the two of them and their friends, Ole and his partner Margaret, decided to make a go of it in the Alaska bush. They were the very personification of the counterculture movement that had just swept through the United States and western Europe (as well as Japan, Australia, and New Zealand) in the late sixties; a movement not yet dead by any means. Bruce later dismissed them as "a bunch of hippies," but I objected. Though it was not an entirely unfair label—the Swedes might have used it to describe themselves—it was a label that carried derogatory connotations in society at large, as if the counterculture was somehow irreverent and irresponsible in its rejection of post-World War II and Vietnam War-era mainstream values. The Swedes were part of a fast-growing, back-to-the-land movement that was seen by some as an outgrowth of the counterculture movement in America, but it had firm roots in Europe as well. This rural relocation trend, as it was referred to in academic circles, was sizeable enough in the sixties and seventies that it could be seen in American demographic statistics, but it was sufficiently notable in Sweden to attract the attention of demographers there, too.[18] Compared to the more heavily populated Sweden, Alaska must indeed have looked like the mother of opportunity for back-to-the-landers.

Many participants in this movement were well intentioned but pragmatically unprepared, and soon had to abandon their dreams for wage-paying jobs. By all appearances, however, the Swedes had taken to the lifestyle and were doing quite well. The charismatic Ole and his girlfriend were planning to marry soon, hoping that it would somehow get them past immigration and enable them to stay on the river.

The four of them had just moved down from their winter cabin on the Charley to spend summer at the old Slaven Roadhouse, to garden and tend their fish wheel on the Yukon. The "wheel" was an ingenious device for netting fish with virtually no expenditure of human energy, save emptying the collection box before it overflowed with the harvest. I first took it to be an aboriginal invention, as fish wheels were used widely by native people in the Interior, but history has it that the wheel was introduced on the Tanana River in 1904 by an Anglo.[19] Constructed like a paddlewheel, it consisted of two wide, gracefully curved, basket-like nets made of birch saplings and woven willow shoots, alternating with two broad, flat, wooden paddles. Baskets and paddles were

Fish wheel on the Yukon River near Coal Creek, 1975.

mounted on a spindle supported by a crude framework that extended out from the riverbank. The river current turned the wheel, dipping one net at a time into the water. Salmon migrating upstream in the muddy waters would swim headlong into the turning net, and as the flowing water brought the net nearer the surface, the paddle behind it would engage the current and drive the basket with its catch up and out of the water to tumble the fish into a holding box. It was a clever contrivance and from what I could see, Ole had not come into the country without considerable building skills. His craftsmanship in assembling this fish wheel from the materials at hand was impressive. Their wheel would free them now to concentrate on processing the fish, which in itself would occupy much of the summer. It also freed them to work their garden, which they had planted next to the cabin. Potatoes, beets, carrots, chard, lettuce, and peas provided welcome relief from their more monotonous winter diet.

Preserving salmon was a critical hedge against a possible scarcity of moose in the fall. Once the salmon run got going, the Swedes could expect to eat their fill and put up as many as seventy-five to one-hundred king salmon for the winter, dried or canned, with some fish weighing fifty, maybe sixty pounds. Heads, guts, and eggs would be fed to the dogs and used as fertilizer for the garden. In late August, when the dog salmon were running, they might see a hundred fish a day fall into their collecting box. A few of these would be eaten as they were caught, some river folks even preferring them to king salmon, but most would go to the dogs. Split and hung on the drying racks, anywhere from one hundred fifty to two hundred salmon per dog would be needed to keep them working through the winter; nearly a thousand fish for five dogs. But the salmon run would not get going full bore this far up the Yukon for another couple of weeks yet, so now was a time for relaxing and visiting. We sat on shore for a while, welcoming the break, and talked with our neighbors over strong boiled coffee.

Our conversation eventually worked around to gold prospecting. Coal Creek, better known for its glittering ore than the name would suggest, was the last major holdout of the gold diggers on this stretch of the Yukon. When the Federal government in 1934 raised the price of gold to thirty-five dollars an ounce, investors got serious about the buried placer deposits of some of the river's larger tributaries. Before the year was out, a well-funded Canadian General by name of A. D. McRae bought up the leases on all the claims of Coal Creek and launched a daring plan for a large-scale strip mining operation. He ordered a huge, floating, bucket-line dredge from San Francisco and had it shipped, dismantled, by steamboat to Skagway, then by rail to Whitehorse, and finally by barge down the Yukon to the mouth of Coal Creek. From that point he hauled the mammoth machine overland on a road slashed through the bush to a promising site, six-and-a-half miles up the creek. There, in a place as remote and wild as any in the country, he bulldozed a lagoon to accept the dredge; and in his quest for a few ingots of gold, he changed the face of the land forever.

The dredge was as alien and frightful looking a creation as ever devised by humankind. A mechanical monster of appalling animation, it was all head and anus. With an ominous conveyor of sixty-two bucket-teeth, each gouging more gravel in a single bite than a sourdough could shovel in an hour, the monster devoured ground ahead of itself at a rate of two hundred truckloads a day, only to defecate all but an infinitesimal fraction of its metallic appetite out the other end. The dredge sat in its own pond and, like a fat river cow, advanced slowly by eating up everything in its path in fourteen-foot gulps, leaving huge droppings behind. Even its tailings—long, slender, segmented mounds of lifeless earth—looked like some ugly larval progeny. To keep its own pond from fouling with its waste, the dredge disgorged a slurry of sediment into Coal Creek at a rate of a thousand gallons a minute, drinking up an equal amount of clear water from upstream to satisfy its thirst.[20]

The floating monster was attended by a hoard of men whose job it was to cut a swath through the trees, and bulldoze the shrub and moss cover to bare ground. The overburden of dark organic soil, one to two feet in depth on average, was then washed into the creek by high-pressure water lines to expose the frozen gravel beneath. With the placer deposits laid naked, the attendants would then drive heavy pipes into the frozen ground and force steam through them in order to thaw the gravel, though if they could clear the land two years ahead of the dredge, the sun would take care of the thawing for them. This stripping operation alone increased the sediment burden of the Yukon by an estimated four million cubic yards. And for every sixty-six cubic yards of gravel thus obtained, the miners could expect one ounce of gold.[21] Deregulation of gold prices in the U.S. in 1972 kept dreams and the aging monster alive, but the dredge was quiet this summer for having run out of stripped gravels.

As talk wound down, my thoughts lingered up on Coal Creek. I wondered what morality could justify raking the land bare, ripping out its guts and leaving the remains heaped along miles of creek with hardly a backward glance. A varied thrush called from somewhere in the distance, its long plaintive note seemingly in mourning. I fingered the gold ring on my left hand. How much responsibility did I bear?

With midnight approaching, we were getting anxious to reach camp. The sun had skimmed below the horizon, just low enough to cast its nightly twilight over the land before beginning its gradual ascent into the next day, but it was enough to remind us of our tiredness. Already ten hours on the river, we were barely making six miles per hour and had twenty-five long miles still to go, so we bid our hosts good luck on

Kandik River base camp.

the summer and pushed our boat back into the main current. Slowly, the fish camp dropped from sight as we made our way upstream, square bow plowing against the eight-miles-per-hour flow of the Yukon. The sediment load of the river seemed to drag incessantly against the bottom of our boat, ever-trying to turn us in the opposite direction. It was cold out on the water and we all fell quiet, as if to direct our energy toward staying warm and helping time along. In an hour or so we passed Sam Creek, then a short while later the confluence of the Charley. We would come back to the Charley someday to investigate its upper drainages, but our business was on the Yukon first. Another hour and Biederman Bluffs came into view on the right bank. A big, sweeping turn of the river, past a couple of cabins on the left bank, and finally up ahead I could see a plume of clear water merging with the Yukon. Eduardo held to the center of the river without slowing, but fixed his attention on a narrow gap in the trees at the upstream end of the plume. We closed the distance quickly, and once past a big sweeper on the right bank, Eduardo turned sharply into the protected channel of the Kandik River.

Away from the powerful current, a backwater slough of the Yukon quietly met the Kandik, forming a large island to the right which provided a buffer between us and the open expanse of the Yukon. The Kandik then swung eastward to more or less parallel the slough, the two forming a narrow, sandy peninsula between them. We turned into the slough and were met with the sight of camp: a solitary canvas wall-tent pitched in the sand, but as pretty a site at that moment as I had ever seen. We cut the motor and coasted aground, and all of us sat for a moment in deafening silence.

III. KANDIK

MOSQUITO WARS + SAGEBRUSH SAGA + REINDEER TO THE
RESCUE + BEAR STORIES I'D NEVER HEARD BEFORE

Before the French fur traders trapped their way northward and westward from the Mississippi, they had already sampled the scourge of mosquitoes—and had a hint of what was to come. When the Jesuit priest Father Poisson visited the French enclave of Arkansas Post in 1727, he found mosquitoes "the greatest torture, compared to which everything else is positively amusing." He wrote, "This little creature has caused more swearing among the French of the Mississippi than has been uttered since the beginning of the world."[1] Father Poisson should be glad he never encountered the Yukon.

Learning to deal with mosquitoes is perhaps the greatest adjustment to the bush an outsider has to make. Many a tough sourdough has folded under their summer rage. Failure should not be taken as too serious a personal weakness, however, for mosquitoes have long conspired against large mammals. They can claim a newborn caribou in a matter of hours, send moose on a crazed rampage for water, drive bear to higher ground, and incite panic in humans lacking sufficient psychological armor. This they accomplish by sheer number and unwavering purpose. There are only three dozen species of mosquito in Alaska, not many compared to the richness of the tropics and the three thousand or so species worldwide, but the diversity gap is more than compensated for by population size. In the Far North, permafrost beneath the thawing surface soil impounds water, and standing water provides reproductive opportunity for mosquitoes. Ample habitat, especially in heavy snowfall years when meltwater is abundant, coupled with an incalculable number of large mammals to satisfy the insect's needs during the breeding season, assures more mosquitoes—billions of them.

If there is anything about mosquitoes to be thankful for, it is that the males do not bite. The female, however, is truly driven in her quest for mammalian blood. She needs hemoglobin to maximize her reproductive potential and will not be deterred by any action on the part of her involuntary donor, save flight or chemical defense. I once came upon a hitchhiker headed in the opposite direction, who was lying on the gravel Denali Highway, propped against his rucksack with a heavy jacket caped over his head and chest. I stopped the car and backed up somewhat apprehensively, and asked if he were all right. He parted his jacket just enough to see me and replied in a heavy foreign accent that yes, he was okay, but just trying to get away from the mosquitoes that were driving him crazy. It was a pathetic sight as, indeed, the mosquitoes were swarming all over the poor fellow. I got out of my car and rummaged through my gear on the back seat for an extra bottle of bug dope, which I left with him. His thanks were animated and profuse, and as I drove on, I thought about the hundred miles of muskeg and tundra that I had just crossed, wondering if he would make it. I recalled someone's having calculated that a human totally bared to mosquitoes in the tundra would lose half their blood within two hours, and contemplated the dedication of the individual who exposed arms and torso long enough to obtain the data necessary for such calculation. I thought, too, about the aboriginal people of the boreal forest, and how they carried smudge pots at all times in the summer, even in the bow of their canoes, with smoldering bracket fungi or wet moss to discourage the mosquitoes.[2]

On the positive side, mosquitoes might be credited with keeping cattle out of the north woods—not that introduction of the latter hadn't been given serious consideration. The early proprietors of the trading post at Fort Yukon actually kept a couple of cows for a summer or two, Murray himself claiming that upwards of a thousand head would be feasible.[3] Even Lieutenant Schwatka speculated on the possibility of grazing cattle on the lush grasses of the Flats, suggesting that with a little judicious drainage and burning of timber and brush, it might well be practical. Later, however, he rescinded the idea, conceding that "according to the general terms of the survival of the fittest and the growth of muscles most used to the detriment of the others, a band of cattle inhabiting this district, in the far future, would be all tail and no body, unless the mosquitoes should experience a change in numbers."[4]

Awakening that first morning at our Kandik River base camp, I laid in my tent in the mounting heat, watching the mosquitoes bumping, probing, clinging to the netting, whining loudly in frustration. They knew I was in there and they were raging mad. A half-dozen of them had found the minute opening at the top of the screen, where the two zippers met, and got in as I slept. By the time I stirred, they were bloated. I dispatched each of them calmly and methodically once I was fully conscious, and then sat a while contemplating. The score was only temporarily in my favor.

My tent was well-used, having seen many field excursions throughout my graduate school years, but it was still in fairly good condition and I felt at home in it, oblivious to its numerous multicolored patches. On the screen were three small holes plugged with the adhesive part of a band-aid, an expeditious fix at one time and still serving its purpose. The tent was a small A-frame with a tight, sand-colored fly that shed water reasonably well. Rated for two people, it was just adequate for myself and all my personal gear and working paraphernalia. In mosquito country, that nylon shell was like a womb, buffering me in one-and-a-half cubic meters of friendly space against the hostilities of the world beyond the mesh screen. It provided an utterly false sense of security, however, for I was also a prisoner in my tent.

As my body began to awaken that morning to other needs, I looked again at the screen and tried to deny myself the inevitable. But the call was not optional. My body, it seemed, was conspiring with the angry mob outside and the moment of reckoning had come.

To the extent that it was possible, we always tried to set up camp on the fresh end of a river bar, where we might benefit from the lack of cover for mosquitoes and whatever breeze the open expanse of water might offer. But for some reason, we always dug the latrine back in the willow brush at the older end of the bar. The rationale for this was lost, I suppose, in some puritan ethic, and though we were all of the same gender and hardly in need of privacy, not one scientist among us ever challenged the logic of putting the latrine smack in the middle of the enemy's camp. Instead, every day we faithfully did our duty in the willows, and no purple heart was harder won. Credit Eduardo, though, for he had been here before and knew the enemy. I hadn't thought again about that case of insect repellent until I sought out the latrine for the first time—and there it was, the whole lot placed strategically next to the coffee can covering the toilet paper. It was worth your life to bare your tender parts out there, so you dropped your pants fast and sprayed everything you cared about. A humbling experience, I came away grateful for small conveniences.

The next few days were busy and fulfilling, as we tackled our field objectives. All scientists, it seems, work for a singular cause, and that is love of their chosen field. The pursuit of esoteric interests is like a drug to the scientist, a mind-stimulant that elevates him above more mundane concerns. When scientists are engrossed in their work, they are happy, and so we enjoyed an amicable atmosphere at our camp on the Kandik.

Garrett spent a day scouting for suitable sites to start his small mammal survey and then got down to work in his usual methodical way. His basic tool was the mouse trap, a modified version known as a "Museum Special," which had a slightly larger bail and softer spring designed to be less damaging to skulls. He would set these out in lines through promising habitat, three traps at a station, each station spaced about thirty feet apart, 120 traps in all. Most of the time he would bait them with a little peanut butter and rolled oats, but occasionally he would not. The olfactory attraction of a previous animals' scent or of old bait on his well-seasoned traps was sometimes enough, and he often placed traps in narrow runways through thick grass or sedges, where the animal could not avoid them. During these long subarctic days, Garrett would check his traps twice every twenty-four hours, as the voles , especially, were active around the clock. While his traps were working for him, Garrett would be off for hours, rooting around on the river shores and sandbars, looking for all manner of animal sign—tracks, scat, clipped vegetation, tufts of fur clinging to stiff spruce branches, combed off as animals passed through—reading it all with practiced skill.

Eduardo had his traps, too: simple ones fashioned from square pieces of plywood, two-and-a-half feet on a side, the entire surface smeared with an impossibly sticky, non-drying adhesive capable of holding anything that landed on it. These he placed systematically in mature spruce stands, tied shoulder-height to the trunks of the largest trees. To ensure that something would land on his boards, he hung a small vial of made-in-the-lab insect pheromones in front of each. The sticky-traps stayed in place throughout the summer near our Kandik River base camp. For more mobile trapping, Eduardo used a Malaise trap, a tent-like affair made entirely of insect netting, but open on one side so that insects could fly into it. Insects hitting the netting would climb upward until they reached the peak of the tent and then move toward one corner or the other, where they would end up in a collection jar containing ethanol to kill and preserve them.

With his traps set out, Eduardo, like Garrett, would then go off scouting, armed with a butterfly net instead of his shotgun, collecting all manner of insects. He found plenty to be excited about, too. In no time, Eduardo discovered tiger beetles in the sandy banks of the Kandik, a significant range extension from the previously known distribution of these ground-dwelling predators. The very next day, he came up with two specimens of a rare cerambycid beetle that he later learned were only the third and fourth specimens ever collected in Alaska.

For much of the time working from our Kandik River base camp, our interest was focused on Kathul Mountain, just a few miles up the Yukon. Here was an entire hillside that appeared an anomaly. Deep in the boggy interior of the boreal forest, Kathul had a distinctly un-Alaskan character. It was like an orphan in a foreign land. Geologically, the mountain was part of the ancient core of North America; the rest of the state having been plastered on via a train of collisions with drifting crustal plates. In its vegetation, though, Kathul Mountain looked more like the steppes of Russia, a remnant of the last great glacial episode when Alaska was part of a mainland biome that stretched uninterrupted all the way across northern Eurasia to England.

What made Kathul Mountain seem so out of place was the dominance of sagebrush on its southern slopes. Fourteen thousand years ago, and for a period of time going back perhaps another ten to fifteen thousand years, sagebrush and grasses were apparently very abundant here, judging from the pollen assemblages found in buried soils and pond-bottom sediments. This was a time when continental glaciation had reached its maximum elsewhere in the northern hemisphere and sea level was at its lowest, exposing an extensive land mass that connected Alaska with Asia. Altogether, this region known as Beringia— which encompassed central Alaska, the land bridge, and eastern Siberia—was large enough to have experienced a very continental climate, with most of its interior remaining quite dry and ice free. Extensive grasslands and sage had covered the landscape.[5]

It was an exciting time in the annals of natural history, for with the spread of grasslands into Alaska, several large herbivores grazed their way into the New World. Most notable among them were the woolly mammoths, who show up on cave paintings as far afield as western France and whose mummified remains turn up now and again in interior Alaska. Following the mammoths and other grazers were hunters: rugged Mongolian stock who many think became the ancestors of all North American Indians.

With the eventual warming of the northern climate and resultant waning of glacial ice, sea level rose, Alaska was cut off from mainland Asia, and the arctic steppe in the New World became relegated to isolated islands. The woolly mammoth and several other associated herbivores of the ancient biome went extinct, though not before migrating all the way to the United States/Mexico borderlands. With the encroachment of boreal forest from the south, so eventually did the steppe vegetation disappear; that is, until Steven Young, my boss at the Center for Northern Studies, went looking for it.

It was Steve who deciphered this Asian connection during field studies the previous summer. Reasoning that the arctic steppe must have been a cold place, he had started his search for a modern analog of the extinct biome on the expansive, treeless plateau lying just to our north. However, finding only tundra vegetation similar to that which covered vast areas of northern Alaska, and convinced that no extensive relicts of the arctic steppe were likely to be found, he returned to the steep river bluffs where he knew sagebrush grew in scattered pockets. It was on the hot, dry, south-facing slopes of Kathul Mountain that he discovered what he had been looking for. There, he found thriving specimens of almost every floristic element represented in the pollen record from ancient soils. Some of the rarest of Alaska plants were growing in abundance among the sages, separated by a thousand miles from their counterparts on the steppes of central Asia—stranded by a drowned land bridge. While Dr. Young was with colleagues in Leningrad this summer, hoping to corroborate his ideas about the remnant plant community from the other end, we were collecting additional field data that might either support or refute his thesis.

So each day now, we headed up the Yukon to the base of Kathul Mountain, where we dropped off Eduardo and Garrett to scour the steep slopes for other relicts of a lost arctic steppe. Bruce, Mark, and I would then take the boat on up to the Nation River, where Bruce wanted to work some fossil-bearing deposits, and Mark and I established some forest sample plots for base-line data collection. It was an enjoyable routine, and I was always eager to pick up Eduardo and Garrett on the way downriver and hear stories of their day. It was a much more difficult routine for the two of them, as they had to bushwhack some distance through dense willows and alders with hoards of mosquitoes, just to reach the start of their climb; and then they were faced with a long, very steep slope, hot and dry, and with scarcely a level place to rest. But they were rewarded with frequent sightings of moose and grizzly from their high vantage point, and came to look forward to their daily trek. Then one day, Eduardo was waiting for us with a grin bigger than life. He had just documented two of the rarest butterflies in Alaska, the Anicia Checkerspot and Chryxus Arctic, spotting several specimens of each among the sages. Two more endemics, it looked like, for the relict sage-grass community. Steven Young would be pleased.

Back in camp that evening, we enjoyed quiet reflection on our day's work. It was an especially still twilight and everything seemed just right. There was little talk in camp, each of us preoccupied in a contented, unrushed way with our own business. Garrett was preparing museum specimens as he usually did in the evenings, sitting on an upside-down bucket at a makeshift table in the cook tent, where he kept a smudge of pyrethrum burning to discourage the mosquitoes. It was a sensible tactic, I thought, since swatting at whining mosquitoes with a half-skinned vole in one hand and scalpel in the other was bound to be messy, one way or the other. Garrett skinned voles with the same fascination and attachment as a philatelist fingering through his stamp collection. And like all collectors, he had his hopes set on the exceptional find. For Garrett, the prize catch this summer would be a singing vole, *Microtus miurus:* an unusual animal for both its high pitched vocalization and its habit of stockpiling grass clippings, much like the pika. He had never seen one, but the diminutive rodent was reputed to be found in dry, rocky habitat such as he'd been trapping on Kathul. To find one would truly be exciting and we all took an interest in the possibility. Garrett's hopes were raised a couple of days earlier, when he had come down from the ridge with a specimen having a conspicuously short tail that seemed to fit the description, but close examination in camp revealed grooved incisors; probably *Synaptomys,* he concluded, the northern bog lemming out of its usual habitat.

Mark, having no research obligation other than to assist me, spent a lot of time writing, and now had gone down to the far end of the river bar where the driftwood piled high, to distance himself from the others. It seemed his thoughts were elsewhere, but he surprised me when he returned after a while with a poem for Garrett—his own ode to the singing voles. Garrett was visibly pleased as he scribbled the lines into his journal while Mark read them again.

Bruce spent his evenings caressing his lode of dull, grey mudstone, holding each fragment in turn close to his thick glasses, studying every minute detail of the rock and the remnant castings of ancient time stamped upon it. Through the tiny window of his hand-lens, a rock in Bruce's hands could be transformed into a primeval forest of an entirely different time and place, a world only he could see while the rest of us imagined. In an admiring way, I was continually fascinated by Bruce's labors.

I stretched out in the sand that evening, leaning against a smooth, silver spruce log as weathered as the land itself, and mused over the scene like a father proudly watching his children. Even the smoke from the dying campfire hung lazily that night, my own mind drifting aimlessly with it.

I thought about the Natives, who once camped along these banks. This was Han country, the former homeland of a small group of people who were culturally similar but otherwise unrelated to the Kutchin downriver, and numbered only three hundred or so when contacted by the first whites. Alexander Murray, who occasionally saw them at his trading post, called them the "Gens de Foux,"[6] which, could be taken variously as wild, unruly, or mad men, though Whymper referred to

Garrett dissecting voles in the cook tent.

the name as a "flattering epithet"[7] The Han, like all indigenous people of the Interior, led a very nomadic existence within loosely established territorial boundaries, moving around in small bands that usually only included extended family, following the availability of unpredictable food resources. It was only after the influx of whites that they accepted a somewhat more settled existence, and one of their permanent villages was established where we were now camped.

The word Kandik evolved from the Han "Kay^{n}dik," meaning "willow creek," the name they gave the location.[8] In time, the settlement became known as "Charley's Village," after Chief Charley, and the river as "Charley Creek," often confused with the Charley River downstream. Lieutenant Schwatka stopped here in 1883 and described Charley's Village as an exact replica of Klat-ol-klin' (Anglicized to Johnny's Village), upriver from Eagle, which he had taken the trouble to document and sketch in great detail. The six houses were constructed of pole-sized logs in which, according to Schwatka, "ventilation seems to be the predominating idea," adding parenthetically "although even this is not developed to a sufficient degree, as judged by one's nose upon entering." The doorways were covered with moose or caribou skin, and the roofs were of skins battened down by spruce poles that projected beyond the ridge line in irregular lengths, often by six or eight feet, giving the whole village "a most bristling appearance." In spite of their airy construction, the interiors were apparently so smoky that the occupants kept close to the floor (according to Schwatka, the air was so thick that it was impossible to stand upright), while the area under the peak of the roof was hung with salmon for smoking. Drying racks by the river's edge kept still more split salmon out of the reach of the sixty or seventy dogs that typically inhabited these Indian settlements.[9] But if their physical structures were lacking in any significant way, the people of Charley's Village apparently made up for it in personal character. "Jo" Ladue, a Canadian voyageur living among the natives at that time, told Lieutenant Schwatka, "[these] are the best natured Indians from here till the Eskimo are met with."[10]

The Han subsisted in large measure on river resources, but historically had access to the Fortymile caribou herd as well, and relied heavily upon the caribou for meat and winter blankets, the fur of this animal being superior to all others in the Interior for warmth. They hunted caribou primarily by means of "surrounds," drift lines made of poles and brush set in the woods that served to funnel the animals into a corral, where they were shot with arrows or snared in openings spaced around the corral.[11]

Caribou were once very plentiful in this area. The indomitable Hudson Stuck, on a cross-country journey from the Tanana to the Yukon, noted in 1912 that the banks of the Charley River, for a distance of fifty miles, had been "trodden hard and solid by innumerable hoofs of caribou."[12] But caribou herds are subject to wide fluctuations in number. By the end of the 19th century, both caribou and Han had dwindled considerably, and for much the same reasons. The land was getting crowded and the hunting pressure increasing.

So scarce had caribou become in this area that by fall of 1897, rumors began circulating in the Lower Forty-Eight that miners on the Yukon were facing a winter of starvation. This prompted one of the wildest schemes for disaster relief ever conjured up by the Feds in Washington. On December 18, 1897 the House of Representatives passed a bill appropriating two hundred thousand dollars from the Treasury and authorizing the Secretary of War to purchase, at his discretion, subsistence stores for the relief of the people along the Yukon (although presumably not including the Han), and to provide the means to distribute such stores. It was also decided that reindeer would be the most appropriate vehicle for transport. On December 23, Dr. Sheldon Jackson, an agent of the Department of Interior, was instructed to go to Scandinavia and purchase five hundred draft animals, complete with sleds, harnesses, drivers, and fodder.[13]

Jackson must have found a bargain, for on February 27, 1898, he landed in New York harbor aboard the steamer Manitoban with 538 reindeer, 418 sleds, 411 harnesses, 113 Scandinavian immigrants, and whatever remained of the 250 tons of reindeer moss (actually lichen) that was purchased for the journey. As soon as a contract could be advertised and awarded for the next leg of their trip, the reindeer were loaded onto box cars and the whole entourage then shipped by rail to Seattle, thence by steamer again to Haines, landing there on March 29, 1898.[14]

It was an extraordinary expedition, and not without its critics. Alfred Brooks of the U.S. Geological Survey in Alaska referred to the reindeer as "government pets," decrying that the ordinary use by miners of horses and dogs to transport supplies in that country was not deemed sufficiently spectacular by government officials. And while the "Santa Claus team," as he dubbed the mission,[15] steamed toward the salvation of its compatriots in the Alaskan deep-freeze, reports were filtering back to Washington that accounts of destitution on the Yukon had been greatly exaggerated. Confirming these reports through returning miners, the Secretary of War called off the relief effort before Jackson reached Haines.[16]

From there, history and the accounting of the reindeer becomes vague. The promoters of the mission attempted to save face by proposing that the animals be used to outfit exploring parties then being organized by both the Geological Survey and the Army. However, long delays and inadequate pasturage in the Haines area resulted in the death of many of the animals. By June, only 144 of the original reindeer remained. The Army deployed a single expedition with these, and in February, 1899, 114 surviving reindeer, guided by 6 Lapp herders, passed by the present-day site of our camp on the Kandik, on their way to Circle City. Whether the 30 missing reindeer died en route or wandered off to mix their genes with the Fortymile caribou, a distinct biological possibility, is unrecorded. At Circle, ownership of the remaining animals was passed from the Department of War to Jackson and the Department of the Interior, and thence into obscurity.[17]

For the Charley Creek Han, none of this, of course, had any meaning. They were already locked in the downward spiral of small-population demographics, with the end in sight. In 1914, the Yukon itself dealt the final blow. During spring break-up, the river flooded its banks and took Charley Village out with the ice. Most of the surviving Han moved down to Circle to take up a new life as best they could.[18] It was an undignified ending, and if any further encouragement were needed to abandon their traditional lifeway, it was provided the following year by an Act of the Second Territorial Legislature. The new government would grant citizenship and all its rights to the Han or any other indigenous person who would swear before a magistrate of the U.S. District Court that he or she "forever renounces all tribal customs and relationships." For verification of their intent, the court required of the Alaskan natives only that they obtain the endorsement of "at least 5 white citizens of the U.S. who have been permanent residents of Alaska for one year."[19]

My thoughts were still with the Han when the canoe slipping past the alders that lined the slough behind us jarred me back to the present. Ironically, it was human encounters in the bush that I was least prepared for, and in three million acres of Alaskan wilderness, we had unexpected company. But if I was concerned, I did not show it. I showed nothing. I just watched incredulously as this person approached, processing the information my eyes were gathering as fast as my brain could work. But the data were not coming fast enough. The green canoe slid toward us as quietly and stealthily as an alligator in a black-water swamp. Not a sound did the figure in it make. His paddle sliced into the water with practiced precision and efficiency, barely rippling the surface. His two dogs sat at absolute attention. We were all on our feet now, Eduardo holding his anxious Lab.

Without hesitation or any sort of acknowledgement, the silent figure came straight on for the beach. The twin huskies lurched awkwardly in the bow as the momentum of the canoe ground it onto the sand.

That our visitor was a white male was my first revelation; not too surprising in this country, but at the same time not readily apparent, for every part of the man's body was heavily covered against mosquitoes. There was not a glimpse of flesh exposed anywhere. His unwashed hair hung straight down in coarse, dark brown strands to the pockets of his soiled khaki shirt, completely veiling his face and neck. Ragged bandanas pinned around his sleeve cuffs hid his hands and extended his arms almost to his knees. Below the knees, his crusty brown denim pants disappeared into high rubber boots. He buffered himself from the chill river air with a salt-rubbed, moose-hide vest, worn flesh-side out.

The man parted his hair and to my surprise was quite gentle-looking up close, peering out through gold wire-rimmed glasses. He offered his hand and said his name was Fred, and that he'd been upriver a ways and smelled our campfire. His face seemed young, late twenties I guessed. He leaned into the canoe as he continued his introduction and came up with a beaver that he had just shot. You could see the satisfied pleasure in his eyes as he held it up by the tail. He knew this was a fine offering. In the bush, this would earn a welcome in anyone's camp.

Eduardo accepted the carcass far more graciously than I might have, for which I was thankful. My senses were still in a state of confusion over what was happening. Fred was not a sight you could welcome unreservedly, and I was not yet sure that I wanted him at my party, so I was glad when Eduardo took the lead.

At our campsite, Fred opened up like a man who had not seen anyone for a long time, carrying on an uninterrupted monologue as he skinned-out the beaver, while the rest of us just sat, listened, and watched. He recounted how he'd gotten careless a week ago and rolled his canoe. Lost his rifle, among other things, so he had to go after this beaver the hard way: with an antique .22 revolver.

"Only the trigger's broke, so you have to hold the hammer back with your thumb and just let 'er go when you've got your shot. Real tricky in the water," he added proudly. "Gotta be patient. Hafta wait till he comes right up close and then shoot accurate, 'cause if you don't kill him dead away, the beaver'll go straight down an' you'll never see 'im again."

He propped a stick against our sitting-log and draped the skin over it, then started cutting up the meat and tossing chunks into a pot Eduardo had offered. On he continued, proffering all kinds of advice on getting along in the bush, mostly with mosquitoes. Fred didn't have money to spend on canned repellents and wouldn't be bothered with bug dope anyway.

"Just don't go an' wash any more than you can help it. Stuff like soaps an' shampoo just attracts 'em. An' deodorant, too. Deodorant's worse of all."

He lived by his word.

Fred was not particularly watchful of the sand on his oily hands and knife, and when a piece of meat missed the pot, he snatched it up, gave it a slap against his pants, and dropped it in. He worked quickly and efficiently, and in no time had every edible scrap of the beaver cooking over the fire. I was getting concerned now, as the etiquette of the situation began to sink in. I would soon be expected to indulge in this meal.

Eduardo contributed an onion and some lentils to the pot, but it didn't help me a lot. An eternity passed while the stew simmered, until Fred mercifully put an end to the wait by declaring dinner ready.

I could probably have just said something lame like, "no thanks, I already ate," but part of me wanted very much to try, and to like, my

"Kandik Fred" skinning a fresh-killed beaver for dinner.

first beaver stew. So I rationalized that beaver served up Bourguignon by a French-accented waiter in finer quarters would undoubtedly be the ultimate dining experience. This was it: haute cuisine of the bush, I told myself. But acculturation is a slow process and I was, after all, brought up in a society that seemed to prefer not knowing where its food comes from. My greatest obstacle was not the oil, or the grit, or the taste (quite good, actually). My problem was mostly in recognizing the parts that I was served. I chewed heartlessly, grumbling to myself something like, "Garrett should be so lucky. He's probably naming everything that goes down."

Over the next few days, I got to know more of Fred and found him quite likable. I wasn't far off about his age. He was twenty nine and had come into the bush five years earlier from Connecticut. He told me that he had made it through high school out East only because nobody wanted to keep him back, lest they have him in the classroom for another year. So they gave him D's and moved him along, and he graduated nearly illiterate. He married young and had kids right away, and kept things going by working construction, pulling all the overtime he could. Then he woke up one morning and asked himself what he was doing. By week's end he was on the road, hitching to Alaska.

He came in the way most did, through Eagle and down the Yukon. Eventually, he established himself a few miles up the Kandik, where he resurrected an old trapper's cabin. "Never slept in it though. Ain't slept under a roof since I came here, not even in winter. No desire to, either." In the wintertime, he told me, he'd just pull a couple caribou hides outside and sleep between them in the snow. From the upper Kandik, two days of walking would put him over the divide and into the Black River drainage. This opened up a huge territory for him to explore. He said he'd once followed the Black all the way to the Porcupine, some two hundred miles, and from there went on down to Fort Yukon and caught a ride on a freight barge back up the Yukon to close the circle.

While a few of the homesteaders on this stretch of the Yukon used square-ended canoes with small six-horsepower outboards, Fred traveled against the river currents by lining his canoe with his two huskies. This was common practice, going back to the Han, before "gas kickers" were brought into the country. The dogs were tethered on long lines attached to the bow and then put on shore to run. All Fred had to do was use his paddle as a rudder, keeping the bow angled slightly into the current. When the dogs came to a steep cut-bank or driftwood snag, Fred would pull in close to shore, whistle his dogs into the canoe, and then paddle as hard as he could across the current to resume lining on the other side. By this means, he could make Eagle in four to six days from his cabin.

Fred stayed around for a week or so and was good company. He never joined us in the field, but would always show up in camp shortly after we returned at the end of the day. He'd sit by our smudge fire late into the evening, habitually spitting on the hot coals, sipping rum from our camp supply, and talking endlessly, almost as if stocking up for the interminable loneliness of winter. We would simply go about our work, and whether we listened, believed, followed his advice or not, didn't seem to matter to him. We were, after all, only temporary visitors. There had been many others floating the river the past few summers, all trying for some experience by which they could say they knew the bush, but they wouldn't know it really—not in one season. So Fred just shared his experiences and enjoyed it. And I was beginning to see him not so much as a recluse from society, but as an aborigine. He had learned what the land had to teach, and in truth, he knew more woods ways than many an Athabaskan adolescent. Fred carried himself softly, seemed genuinely comfortable in the bush; at home, nothing to prove.

I was also beginning to feel a connection to Fred that I couldn't fully explain. There were some rough parallels in our lives: points of dissatisfaction, of not knowing where we were going, of being drawn towards the North and an independent life. In the end, though, my solutions were quite different from his. After a strong start in college I lost my focus, drifted aimlessly amid the ivy halls, and was soon dismissed for academic failure. Without a future, I also married young, at about the same time as Fred, and wandered into the canyons of Boston looking for the light, only to find myself lost again, this time on the seventh floor of State Street Bank and Trust Company. Training for a position in the Trust Department was about as blind an alley as I had ever run into. Wondering where to go next, I was sitting in my apartment one night watching a National Geographic program on television about a bunch of field scientists working out of a tent camp in some godforsaken place, and it hit me like a revelation. That's what I want to do. I never bothered to ask if those guys were making any money; having a direction was all that mattered.

So just about the time Fred struck out for Alaska, I set off on the road to my own dream. I was carrying a considerably heavier load this time around, however. Readmitted to the university on academic probation, I now had a one-year-old son and my wife was ill. I held things together by washing glassware in a chemistry lab, laboring in the stifling, sulfur-infused greenhouses of a commercial rose grower, and picking apples on weekends. I got my son up before dawn, cleaned, fed, and played with him until he fell asleep again, ran off to classes, and came home from school as soon as I could to take him for a walk in the stroller while I did my required reading on my feet. A few months later my daughter was born. Once I had the dream, though, I wouldn't let go.

I thought now about Fred, and felt the pain of leaving family for an uncertain future. I felt his pain when he told me of hitching back to Connecticut, after being in the bush for a couple of years, to see his kids and not being allowed in the door by their mother. But men and women have been driven by voices louder than family, and I didn't know enough about Fred's life in Connecticut to make any judgment. I recalled Tolstoy's stunning opening line in Anna Karenina: "All happy families are like one another; each unhappy family is unhappy in its own way." I knew only that the drive to explore, to disperse, to find new ground was built into the mammalian brain, and had been just as important to the long-term adaptation and survival of the species as growth and reproduction. The forces that brought Fred here were no different than the forces that attracted Ole and his Swedish and American companions, and the scattering of others now homesteading on the river. Call it romanticism if you like, but romanticism here was not about merriment and music making, bare skin and sun seeking, raising wine glasses at the end of days that blend one into the next. Romanticism in the bush was about freedom, independence, living off the land, surviving by one's primordial skills and not by out-competing others for jobs and promotions. Fred was clearly not driven by a desire for company of the opposite sex, either. He admitted he wished he had a woman, but lamented in the same breath that "good bush women are hard to find." Given that a successful partnership anywhere starts with an appropriate personality match, which in turn requires a large enough pool of men and women from which to find that match, and given that I hadn't noticed a lot of single women running around out here craving a life of isolation, it seemed no wonder that Fred hadn't found "a good bush woman."

It was interesting to me now that our divergent paths should come together in the Alaskan bush. Fred had failed in a system designed to educate the masses, but survived in the bush with a natural intelligence, understanding what was essential. I wondered how the masses would fare on his ground, but I already knew the answer. He caused me to think about my own education, about what I had been taught and what I might someday pass on to others. I questioned how much understanding had been imparted for all the information crammed into our heads. When Fred talked about wetlands as "high energy" areas, he understood something that cannot be learned in school. He had never measured primary productivity, nor followed the mathematics of energy flow from one trophic level to another, accounting for the efficiencies of transfer from plant to herbivore, herbivore to carnivore, carnivore to detritivore. He simply knew that life concentrated in wetlands because food was abundant, and that he could support himself better by tapping into that abundance than by stalking a caribou on dry ground. He didn't know that the blueberries gave him back 650 kilocalories for every hour he spent gathering them. He just knew that he felt a lot better when he had a supply of them put up with sugar for winter. He didn't know how efficiently beaver and muskrat converted the chemical energy of plants into animal protein, but he understood that beaver and muskrat gave him meat to slake his hunger and bought him flour and honey besides, when he took their winter furs to Eagle. He hadn't read that the energy yield of moose varied seasonally, from 8,200 to 95,600 kilocalories per hour invested in their capture. He just told me that if he shot one fair-sized moose after freeze-up, it would get him a good way through winter, and that if he didn't, he would have a hard time of it. Fred brought fresh relevancy to my science, to the numbers and theory of ecology. It was a relevancy that had somehow been lost from human considerations in a society of industrialized agriculture, where we were willing to pay more in energy costs to produce and distribute food than we got back at the table.

When we got around to talking about bears, as inevitably happens in this country, the tone of the conversation changed. There is definitely an aura about bears that sets them apart from all other animals in the

North. In the spiritual ideology of the Athabaskans, the bear is among the most powerful of all souls in the natural world, a status perhaps not unrelated to its physical strength and human-like qualities. Numerous rituals and taboos, especially for women, are connected with the killing and utilization of bear in deference to their powerful spirits, and violation of these taboos can be met with serious misfortune.[20] But even among Anglos unexposed to native tradition and spiritual connections with the natural world, there is something almost mystical about the bear, something that seems to challenge basic human supremacy. While it may have a lot to do with perceived danger, bears are not the greatest threat to human safety in the wild, nor are they inherently more dangerous than other mammals larger than ourselves. A big grizzly doesn't stack up in size against the largest moose,[21] and woodsmen accord the greatest of respect to the latter, knowing full well that a crazed or belligerent moose can be as formidable an obstacle as any. A bull moose in rut or a cow with calf can be extremely dangerous, provoked by no more than the snapping of branches that innocently mimics the thrashing of antlers in the brush. In the wintertime, hampered by deep snows, moose often get onto packed trails where they will challenge anything that obstructs their path, charging wildly at dog teams or snowmobiles. Still, there seems to be an essential difference: with even a bad-tempered moose, there is never any doubt as to who is at the top of the food chain. With bears, there is just enough doubt to keep a person looking over his shoulder.

Powerful spirits notwithstanding, bear hunting ranks high among the subsistence pursuits of most folks on the river. Among indigenous people, the favored means of taking bear was by den hunting. Bears are in their best condition at the time of denning, when they often will yield enough fat to render four or five gallons of lard, a significant bonus to the meat. Taking a bear from its den, however, requires the utmost of skill and not an insignificant amount of courage.

A den is located by reading subtle signs in the snow for past activity: broken branches of blueberry, often gathered by the bear for bedding; traces of moss or grass taken into the den and used to plug the entrance; old track impressions under more recent snow. Once found, the den's occupancy is confirmed by locating the concealed opening and probing inside. Many hunters these days prefer to shoot the bear in the den and then wrestle it out, but for generations past, and occasionally even today, Native hunters would take the bear with an axe instead. They would chop an opening into the top of the den just big enough to give them a clean swing at the bear, yet small enough to slow the bear if it erupted through the hole. The advantage cited for this method was that it saved ammunition. Fred pondered that for a minute and said he hoped he'd never need a bear or the ammunition that bad.

An old-timer named James Huntington, from down on the Koyukuk River, once described to a writer his experience of unintentionally taking a bear like this. He had followed bear signs to a den, hoping for an easy kill, and proceeded to clear snow from all around it, looking for the entrance. When he still couldn't locate it, he started poking around with his rifle stock until he hit a soft spot. "The rifle went clear through," he recounted, "but still I couldn't feel any bear in there, the way you were supposed to." He started pulling grass and leaves out of the hole and then, "lying stretched out on my stomach, I shoved the rifle in as far as I could, and when I still felt nothing, I shoved my arm all the way in, too. Then I felt a bear all right—a ruffled and cranky bear who growled once, then slammed a paw down on the rifle and jerked it right out of my hand. I yanked my arm out of there as if I'd touched fire." That left Huntington with his .30-.30 inside the den and no choice but to go after the bear with his axe, the way his uncle "Hog River Johnny" had taught him.[22] The first time I heard that story, I couldn't help but think about a bear researcher named Rogers from Minnesota, who would crawl inside occupied dens, armed only with temperature sensors and syringes, and stick and probe the bear, trying to figure out physiologically whether bears in winter were truly hibernating or just in a state of deep sleep.[23] I wondered what Huntington would have thought about that.

Early ethnographic accounts of the Han describe yet another strategy for taking bear. A hunter, armed only with a lance or spear, attracts a bear by making raven-like noises, since ravens commonly scavenge on carcasses that would be attractive to a bear as well. As the bear is provoked to charge, often with the help of two others on either side of the bear, the hunter stands firm and attempts to jab his spear into the bear just above the breastbone, at the same time thrusting the end of the spear to the ground so that the bear impales itself under its own weight.[24] It was an innate skill, I supposed, for I couldn't imagine learning by mistakes in that business.

As fate would have it, bears were giving trouble on the Yukon this summer—mostly the black bears, which are a certain nuisance once they catch on to you—and already two had been shot on the Yukon, at the mouth of the Nation River, for tearing up the tents of a party of field geologists. While the grizzly may have a reputation for being both the meanest and also the most curious, a troublesome combination for sure, black bears are notoriously unpredictable and this is cause enough to be wary of their presence. Once a black bear gets into camp and discovers food, there is no getting rid of it short of shooting it. And to venture into the bush without a bear gun is to elicit the sternest of looks from the locals, who view such recklessness as an act of extreme ignorance. That may be so, and we had our protection—two shotguns and a .444—but there is another side of the issue. Firearms can be a liability in the hands of the inexperienced, and when on our first day I decided we should have some target practice, I discovered where the greatest danger lies. I appointed Eduardo guardian for the time being, and put the other two guns harmlessly away.

A week passed by easily on the Kandik. With one day melding seamlessly into the next, any semblance of a schedule was soon lost. Except for our collecting trips upriver, we worked, napped, fixed meals, did camp chores pretty much as we wanted. For an occasional change in scenery, we took the boat out on short forays to explore nearby sloughs and tributaries.

One lazy morning, three of us drifted downriver to see if we could find the remnants of a roadhouse that Hudson Stuck had written about in his journals. He described the location as ". . . just below the mouth of Charley Creek [a synonym for the Kandik] where the river takes a sharp bend to the right and then to the left."[25] Roald Amundsen, the famed Norwegian explorer, had passed through this area about the same time as Stuck, and provided a colorful description of roadhouses on the trail between Circle and Eagle, so I was curious to see the setting. (Amundsen was traveling overland from his ship *Gjöa*, anchored at Herschel Island in MacKenzie Bay, to the telegraph station at Fort Egbert in order to send news back to Norway. He must have had something important to say, for it took him thirty-eight days by dogsled to reach Eagle.[26])

According to Amundsen, these "small log huts" that provided food and lodging to travelers were "situated along the river at intervals of about twenty miles, and generally consist[ed] of three rooms, the room for the guests, the kitchen, and a little room for the proprietor." All arrivals, he noted, were packed into the first room. "Those who have not their own beds with them must share with another, but people in these parts, after traveling all day and arriving very tired are not very particular." He added, "for us, who had come from the northern regions, these 'hotels' were perfect wonders of comfort and elegance." Amundsen noted that they were also very expensive— "the sleeping place, whether you slept alone or shared it, cost a dollar" —and he complained that "the prices of everything are exceptionally high in Alaska, and, when gold is discovered in the neighbourhood, they go up by leaps and bounds."[27] That was the winter of 1905-1906.

The twenty miles or so separating road houses, by Amundsen's account, may have represented an average day's travel in those times, or it may have been considered the safe distance between competing accommodations. As it was, the mere nine miles separating the Charley Creek roadhouse from an upriver rival on Washington Creek was apparently not sufficient to keep the peace on this stretch of the Yukon. Hudson Stuck, who traveled this reach frequently during the first decade of the 1900s, tells of the woman proprietor at Washington Creek taking shots with a rifle at people who went along the river trail in winter without stopping at her roadhouse. "I cannot speak of this from personal experience," he acknowledged, "though I well remember the intimidating and cajoling placards she posted on the trail." Her story is being told on the river to this day, though I have yet to find a name for this woman. [28]

We were luckless that morning in locating any remains of the Charley Creek roadhouse, but as Stuck noted, "this part of the river is much subject to flood, owing to the configuration of the channel, which lends itself to the jamming of the ice, and this particular road-house keeper is not infrequently camped upon the roof of his road-house during break-up time."[29] Perhaps the establishment was rafted down the Yukon the same time ice took Charley's Village out.

Further downstream, near the mouth of Charley River, we ran into Fred and another boat beached on a stretch of sandy shoreline. Fred's green canoe was easily recognizable from some distance, but the other, a square-ended, flat-bottom boat like our own except wooden and painted bright red, was not one we had seen on the river before. It was fairly well-packed with bundles of white canvas, stovepipe, and miscellaneous poles, shovels and old barrels, one serving as an upright gasoline tank. The operator, in a light-colored, woolen plaid shirt, hunting cap and sunglasses, stood with one foot on the gunwale talking, while another fellow, considerably older and with a slight forward list, with white frizzy hair and wearing a dark wool jacket, stood on shore. Fred sat in the stern of his canoe, bare-armed for the first time since we had met him, in an olive-colored t-shirt. We cut our motor to an idle and let the current carry us around them before pulling upstream to beach beside them.

Fred introduced us to the fellow in the boat, who was also named Fred. As was typical on the river, he didn't offer a last name. Fred-the-latter packed a massive .44 magnum handgun on his hip and apparently it wasn't reserved just for bear protection. I learned later that this Fred had something of a quick temper and a general disregard for problem-resolution by arbitration. "Crazy Fred" was the name by which he had become known on the river. Story has it that Crazy Fred once had a disagreement with a guy in his boat and made him get out on an island in the Yukon. The hapless soul was marooned for many days before someone picked him up, apparently keeping himself occupied by catching and eating bank swallows that were nesting on the island.

The fellow standing on shore was Morris, a Native originally from Circle, with quite a few years on Fred. He held a white plastic teacup from which he sipped often, refilling it part way through our visit. It took me awhile to realize the tea was Jim Beam from a stash in Crazy Fred's boat.

Morris, in his early 70's, was well known on the river. He and his half-brother Silas had trapped the area around the Kandik as far back as the 1930s, and built the cabin that Fred was now using. The two brothers had earned a reputation up and down the river for their woods skills. Covering separate routes, they would regularly catch between seventy and eighty marten apiece, a feat requiring more than a little

Social gathering on the Yukon.

savvy with their trap sets. Marten are solitary animals, not caring much for each other's company except during a short breeding season, and travel great distances when hunting. When times are good, marten might number two to three animals in a square mile. It can get better than that, but it can get a lot worse, too. On average, then, each one of the brothers would have to range over a territory of about sixty square miles if they hoped to catch seventy animals, assuming that each one's take is roughly half the population of martens in their territory.[30] Covering that much ground on snowshoes was no small feat, and for all their effort, the seventy pelts might bring them five hundred dollars apiece.[31]

Morris had spent most of his life in this area, but a few years ago moved from Circle to Nenana. He had come back to the Kandik area for the summer, and now he and Crazy Fred were looking to set up a fish camp—the reason for all the gear in their boat. Morris was animated and talkative. He had a brightness about him, an enthusiasm for life at that particular moment that was notable. One had the impression that if the Saturday night dances were still held in Circle City, he'd be out there kicking up his heels with every woman on the floor. Or maybe it was just the elixir of bourbon and new faces, a rare social event in the bush on a fine summer morning. Regardless, I took a liking to Morris and wished I could know him better, but that was the last I ever saw of him.

Parting company on the Yukon, we motored a hundred yards up the Charley River to satisfy our curiosity before heading back to camp. The Charley turns lazy at the end of its three-thousand-foot drop, wandering for its last few miles through flat peatlands, its clear water picking up tannins from decaying vegetation to turn lightly tea-colored. We tapped into that lethargic mood perfectly, barely maintaining forward motion against the river's now-gentle current. Puttering slowly around a bend, we encountered a young bull moose standing fully in the open on a slip-off bar, its rich, dark, chestnut-colored coat shining in the sunlight, a splendid representative of the species. Startled by our sudden presence, he raised his head and stared at us just long enough for Bruce to snap off a picture, and then moved off quickly, although not in a panic. This was our closest look yet at the regal but odd animal, and I couldn't help but think of Lieutenant Schwatka's many amusing, though not inaccurate descriptions of the moose upon his own first sightings in these parts: "The great palmated horns above, the broad 'throat-latch' before, combined with the huge nose and powerful shoulders, make one think that this animal might tilt forward on his head from sheer gravity, so little is there apparently at the other end to counter-balance these masses."[32]

As the moose passed under a low, sweeping spruce branch, a cloud of mosquitoes rose in unison from its back and then settled down again as the branch cleared, like dust behind a speeding car on a gravel road.

IV. NATION

CIRCLE'S UNWELCOME COMMITTEE ✦ A RIVER RUNNING
SLUSH ✦ PYGMY SHREWS AND GIANT CLUB MOSS ✦
PERMAFROST PERMUTATIONS

It was after midnight when Mark and I pushed our boat off the sand bar and drifted out into the Yukon's rolling current. We were on a milk run, headed for Circle to re-supply, and it made good sense to leave camp in the dead of night.

Nighttime was good river time. We were close enough to the Arctic Circle that the summer nights were never totally dark (the founders of Circle City thought they were right on, hence the name, but they missed it by half a degree of latitude). Having sufficient light to see at all hours was a freedom I hadn't thought much about until that first long trip upriver; there wasn't anything we couldn't do in the middle of the night if we had a mind to. And in the dim, lavender-gray light of the midnight sky, the wildlife that we encountered on the river seemed less reactive to our presence, carrying on their nocturnal business uninterrupted while we observed. It was like viewing the world through a one-way mirror. The night was no longer secret, and it became my favorite time. I was surprised, too, at how little sleep I seemed to need in the absence of darkness.

With an empty boat now and the river current in our favor, we figured we would make it to the village in something under four hours, barring any problems, arriving while most everyone was still asleep. Frank was an early riser and would open the trading post for us when we got there so we'd get a good jump on the day, rounding up our supplies and pumping gas while it was still quiet. That should put us back on the river by noon, and with any luck, we'd arrive in camp before the following midnight.

It wasn't until we left our boat tied up in Circle that we ran into trouble. We had finished our chores in good time; our groceries were put up in boxes ready to be taken down to the boat and Mark had just capped the last gas can when the apple-red Air North passed loudly over us on its landing approach. We would start loading as the mail was taken in and sorted, then check to see if we had any letters and be on our way. But someone in the village had a different idea.

As I lugged the first two cans of gasoline down to the boat, I could tell that something was amiss. Things just didn't look the same as when we had left. To guard against theft we had taken everything that was unattached up to the post with us, including life vests and our spare fifteen horsepower outboard motor. But someone had messed with the boat and it didn't take more than a second to notice that our gas line had been cut.

My immediate sense was that the party responsible had treated us gently, their intentions clearly not overly malicious. If they had truly wanted to incapacitate us they could have cut the line in several places, or left us with no gas line at all. And worse pranks were easily imaginable. Instead, however, they cut the line close to the intake port on the engine so that repairing the damage required only loosening a clamp with a screwdriver and removing the severed piece of hose, slipping the new end onto the male fitting, and retightening the clamp. For the minor inconvenience, whoever had done this might just as well have left us an intimidating note instead. I looked quickly around, saw no other damage, and began refitting the cut line with no more feeling than if I were replacing a broken shear pin on the propeller.

The message said clearly "we don't want you here," yet I found myself strangely unemotional over the act and even somewhat sympathetic toward the vandals. These people were intensely independent and resented any outside interference with their lives, as I supposed I would. Most stayed on here—or came into the country in the first place—because they preferred a life of self regulation, and the prospect of a national park or preserve of any sort in their backyard was not something they welcomed. Fred had expressed the same sentiment to me earlier. It was not with any intent to break or escape the law that these people sought refuge in the bush—they were generally reasonable and peaceful folks—but they had their own codes too, and were particularly antagonistic toward attempts by government to impose restrictions on their hunting and fishing activities. If a person living in the bush needed a moose or bear, they would shoot one when the meat was right and fall freeze-up permitted its keeping—not when legislators in Washington or Juneau said it was okay. And if the moose happened to be a cow they would take it anyway, as long as it was fat enough. The river people knew they were harvesting game illegally, but felt they had no choice and that there was nothing immoral about it. And they justified their actions with an ethic that demanded complete utilization of the animal. A moose was consumed nose, intestines, and bone marrow, and if the hide wasn't tanned for clothing it was used raw for a toboggan bag, babiche, or bait for trapping. The only immorality in their eyes was in the government's looking at wildlife as a source of revenue. Regulation of hunting for outsiders whose only real need was to satisfy their egos was good business elsewhere, they believed, but how the state could impose a season and bag limit on someone to whom hunting and fishing was a matter of living, not a sport, was beyond the understanding and patience of these people.

It was hard not to feel some sympathy. The regulation of sport hunting and fishing had become necessary almost everywhere people settled in number because of the typically human tendency to over-harvest (as wildlife managers put it politely). Witness the early extirpation of beaver in New England, of buffalo across the plains, of grizzly bear in the southern Rockies, of sea otters in Alaska. The problem in most of these cases was one of relative numbers, hunter versus hunted, and of increasing hunter efficiency. A given landscape can accommodate only so many herbivores, depending on how productive their food base is, and the herbivores can feed a limited number of carnivores because ninety percent of the plant energy ingested by the herbivores is lost through respiration and doesn't transfer up the line. Thus, the land has its carrying capacity, an upper limit to the number of individuals at any level in the energy pyramid that it can support. Left alone, it is a self-regulating system. When herbivores out-graze an area, they either move or die off from starvation. If predators over-run their prey, their numbers crash. When the miners came into the country they mobbed the land, exceeded its carrying capacity, compensated for over-harvesting by importing food through the traders, and then moved on. Now sport hunters are breaking the basic ecological rules because their primary source of energy is not derived from the land they hunt. It comes from other foods, produced in other places.

But the homesteaders seem to know what the carrying capacity of the land is. Maybe it's just territorialism—they call it breathing room—but the effect is the same. They know how much land they need to support themselves and are not so inclined to over-harvest anything, if only because they have limited capability and no way of preserving or exporting the excesses. For all the accounts of homesteaders living up and down the river over the years, traces of their past are scarce—hardly the case with mining claims. Old cabins melt into the ground, forest clearings grow back, moss quickly covers old trails, animals reproduce and redistribute themselves across the landscape.

Nonetheless, subsistence use of Federal land was becoming a difficult issue. It was generally acknowledged that the homesteaders succeeding the hit-and-run gold prospectors, returned some semblance of stability to the woods economy and social structure. Often intermarrying with natives, they established much closer ties to each other and to the land—a harmony that seems to have persisted through time and the slow turnover of faces, almost as if by some form of natural selection. And the law assisted the homesteaders, providing for their occupation of unappropriated public land in the same spirit that it allowed miners to stake their claims—to encourage development of the region. The Homestead Act of 1862 enabled anyone over twenty-one years of age to apply for title to a parcel of land for the purpose of "settlement, residence, and cultivation." "Cultivation," as it was originally intended, might have been the sticky point in the law here for those who weren't developing the land, except that every homesteader on the river took advantage of the long summer days to "cultivate" all manner of vegetables, mostly greens to break their gustatory boredom and root crops for winter. But the real problem was that the present occupants—and most of their predecessors—never bothered to apply for title under the Act. In the interim, Federal interests had changed and there were increasing murmurs in Washington about repealing the Homestead Act.[1] The word "trespassers" was being used more frequently, and now Fred and a handful of others worried that their right to remain on the land might be challenged with a government takeover of management. Hence the mistrust in Circle of our motives.

These were nagging questions and my thoughts lingered over them as we motored upriver. On the way we met a barge coming downstream, loaded with a small, rusted Caterpillar bulldozer—another miner set on reworking an old claim, stirring up dreams that would not die. The miner was legally entitled.

With our field studies on Kathul Mountain finished, we moved up to the Nation River for a few days, taking only our personal tents and equipment, and a little cookware. The move afforded Garrett and Eduardo an opportunity to survey the area and gave Bruce more time to pursue several promising outcroppings that he had located earlier. I, too, had found a prime white spruce stand on a steep hillside that I wanted to study in greater detail before heading off to my treeline studies in another week. It was a slow trip this time, with all five of us plus gear and more than a hundred gallons of extra gasoline, and when we reached the Nation we found ourselves sitting too low in the water to negotiate the mouth of the river. The water level had dropped steadily over the past several days for lack of rain in the high country.

Looking for any good reason not to offload our cargo and portage up the river, we beached the boat and bushwhacked a short distance through the thick willow and alder to inspect a cabin that was situated in a spruce stand at the base of a prominent bluff. I had wanted to see the cabin anyway, and Eduardo thought it might be easier to put up there than to pitch our tents—and safer besides, since this is where the party of geologists ran into trouble with bears.

The cabin was engulfed in tall weeds, looking as if no one had been there for some time. The lowest course of logs seemed to be sinking into the ground, requiring us to stoop through the doorway. Inside was bare and dingy, the low ceiling giving me a feeling of claustrophobia. A small window on one side cast a pallid light on a rough table standing beside a rusting barrel stove on the dirt floor. A makeshift cupboard with no doors hung crookedly to the side of the window. The only seating was an old steel cot with badly stretched springs, many held in place with twine. The cabin was built, as story has it, by a Norwegian immigrant

Eduardo with "Moose," above the mouth of the Nation River

named Chris Nelson, an acknowledged loner who apparently became so talkative whenever he encountered another human being that he earned the nickname "Phonograph." Had he and Fred ever crossed paths, I mused, they might have sounded something like a thirty-three rpm record played at forty-five. Story also has it that Phonograph was found frozen to death here in the spring of 1949.[2] I found it depressingly dank and gloomy inside, but tried to imagine how it might feel on a dark and snapping-cold winter night when the stove was creaking with heat and the candles on the table danced with a light as cheery as the pale yellow mid-day sun. A nice image, but it didn't help much at the moment. We decided to pull our boat as far as we could into a backwater slough where there was a gravel bar suitable for camp. There we unloaded gear and cleared away rocks and scraped out a place to call home for a while.

This had been a busy place during the gold rush days. Almost directly opposite the Nation River, the Fourth of July Creek trickled into the

Negotiating receding water levels on the Nation River. Photo by Ed Holsten.

Yukon, and for the final two years of the nineteenth century, this stream was the center of attention in the area. The Fourth of July Mine, eight miles up from the Yukon, eventually proved to be a large producer. More than fifty people worked the site, and over one hundred claims were filed on the creek. The sourdoughs organized, drew up laws regulating mining activity, and things were looking good for a while. On the left bank of the Yukon, Nation City was founded and soon became an important steamboat stop and supply center for the area.[3]

Bituminous coal was discovered a short distance up the Nation River and in 1897 (or '98) the Alaska Commercial Company began operations to supply coal to the steamboats. Yukon steamers were burning more firewood in a day than a household in Circle consumed all winter, so the extraction of coal here took considerable pressure off the wood supplies. The company sledded about two-thousand tons of coal to the banks of the Yukon before history repeated itself. In 1902 the Fairbanks gold strike emptied Nation City, and the abandoned coal mines soon caved in.[4]

A few foundations and some old machinery are about all that remain of the town, but we encountered another interesting artifact not far from the mouth of the Nation. Returning from a hike upriver one day, we came across a two-seat Piper Cub on skis in the middle of a dense thicket of spruce. The wings and struts had been removed and stacked upright and the engine was hung from two trees. One wing was crumpled a bit, but otherwise the plane looked to be in reasonably good shape. It wasn't until I caught up with Dave Evans and Brad Snow years later that I learned the story—a story with more twists and turns than the river itself; a story that began several months before the ski-plane ended up in the woods.

Dave and his partner Sage, along with Brad and his wife Lilly, had come to the area only a year before our expedition. Finding the Nation River unclaimed at the time, they settled on its banks and went about building themselves a couple of cabins, Dave and Sage's three miles up and Brad and Lilly's a half-day's walk further. All four proved capable with axe and saw and learned the basics of homemaking in the bush quickly. In a month's time, both cabins were finished and comfortable. Then one early-October morning Brad and Lilly were eating a leisurely breakfast, gazing through their open door, when a large bull moose wandered into the river on the opposite shore and waded across to an exposed gravel bar in the middle of the river. It was almost too easy. Brad reached for his rifle and put the moose down right there.

That's when life in the bush started getting a little more complicated for Brad and Lilly. With a thousand pounds of moose now lying out on the river bar, they went to work, Brad skinning the moose and cutting it up, and Lilly carrying the windfall, part and parcel, trip after trip, the rest of the way across the river in their canoe, to the cabin. It was good work and good timing, as the meat would stay frozen now, but in the process of manhandling the half-ton animal, Brad apparently did some damage to himself. By mid-October, two weeks after shooting the moose and still working on it, he was experiencing pain on his right side that was getting worse, not better. Brad was becoming convinced that he had appendicitis. He and Lilly finally walked down to Dave and Sage's cabin to talk over the situation and there they decided that medical attention was in order.

It was a tough decision. Mid-October in that country is not a good time to have to go anywhere. The creeks and bogs were beginning to freeze, but nowhere was the ice thick enough yet, or the snow deep enough, to allow easy overland trekking. On the rivers, flowing ice made travel difficult and dangerous, and landing a plane anywhere in the bush now was questionable. A slow-flying Piper Cub might make it in and out on skis, if there were room enough to land on high ground, but it would be tricky at best and disastrous at worst. So the next morning Dave and Brad did the only thing they could. They slipped their nineteen-foot aluminum canoe, with its square stern and six-horse motor, into the slushy current of the Nation River and started the long journey to Eagle in hopes of catching the mail plane there and flying Brad to the hospital in Fairbanks. Without too much difficulty, the two men were able to negotiate their canoe the three miles down to the Yukon, but once there they got their first lesson in river freeze-up.

River freeze-up is an event of major proportion, and the Yukon is a river of major proportion. Large sheets of ice, not thick yet but the size of small skating rinks, were forming in the quieter eddies of the river and slowly, like large rafts, slipping out of their moorings and into the main current. There the rafts would break up into smaller plates and slowly spin as they floated downstream, bumping and grinding against each other as they turned until they were gradually worn into large circular pieces. On their top side and wrapping completely around their perimeter, a raised rim would form through the freezing of slush that was pushed up onto the edge of the ice. This gave them the appearance of a large skillet without a handle, from which the logical name "ice pan" was derived.[5] There wasn't any way now that Dave and Brad could navigate upstream against this flow of ice coming down.

A shelf of ice was also forming along the shores of the river and the only option the men had was to get onto this shelf and pull their canoe against the current. By attaching a line to the bow and stern, and holding onto this line about a third of the way back from the bow, it was possible to adjust the amount of pressure against the bow to keep the canoe tracking in the current, away from the shore. So Dave and Brad started upriver on the shelf ice, Brad walking ahead and probing the new ice with a pole while Dave pulled the canoe. They walked long into that first night before making camp in the snow and darkness. They put up a simple tarp lean-to and built a fire in front of it. They made a meal of canned beans and Lilly's fresh cranberry bread. They slept in their clothes, awoke early and walked for a second day, then a third. Where the shelf ice extended out over the quieter water behind sand spits, they walked gingerly along its outer edge, the ice sheet wavering beneath their feet. The weather warmed on the third day and the pan-ice thinned a little. Knowing they would have to cross the Yukon sooner or later, they decided now was the time, so they pulled their canoe to shore, got in, and slipped out into the current, navigating with paddle and motor between the bruising ice pans, keeping their nose upstream while struggling to stay even with the current. They reached the left bank safely and resumed lining, making it to the mouth of the Seventymile River by late evening.

At Seventymile, they hauled their canoe up onto shore, tied it securely, and struck out for Terry McMullin's cabin a half-hour upriver. Terry had a powerboat and the men wondered if he might still be able to negotiate the Yukon ice and get them up to Eagle faster. When Terry hiked back down with Dave and Brad to look at the Yukon, he just shook his head and said "nope." The best he could do was to show the men a trail that climbed up and over the height of land, shortcutting two large bends in the Yukon and saving them many more miles of uncertain shelf ice. Dave and Brad stayed the night at Terry's and struck out on the trail in the morning, through a foot of snow. They reached Eagle by nightfall on the fourth day.

In Eagle, Brad consulted a rural nurse who reaffirmed his need to see a doctor in Fairbanks, so Brad made arrangements to fly out on the mail plane the next morning. Dave started back down the trail to Seventymile and Terry's cabin, and a day later pushed his canoe into the slushy Yukon current and floated with the pan ice back to the mouth of the Nation River. By the time he got to the Nation, his canoe was three times its normal weight and Dave had to muscle it ashore and knock large chunks of agglomerated ice off it before dragging it the three miles back up the Nation. Six hard days after leaving his cabin, Dave was back home.

Then came the long wait. Before leaving Eagle, Brad had made arrangements with local storekeeper and pilot, Dave McCall, to fly out to the Nation River to pick up Lilly and take her to Fairbanks to join him. McCall would soon be fitting his plane with skis and was expecting to fly to Fairbanks twice a week to provision his store. He would be happy to stop by the Nation River to pick up Lilly; all Brad needed to do was send him the word from Fairbanks. Back on the Nation, Dave, Sage, and Lilly went to work packing a thousand-foot-long airstrip with their snowshoes, and keeping it packed every time it snowed.

The wait for Dave McCall dragged on, days into weeks. It was mid-November and still no pick-up for Lilly, and no word of Brad's condition. Finally, sometime after Thanksgiving, McCall showed up. The makeshift runway was tight, but McCall got in and out cleanly. Lilly joined Brad, who had suffered only badly torn muscles from wrestling with the moose, and the two stayed in Fairbanks long enough to earn back their plane fare (pipeline jobs could be had just by showing up) and pay off the doctor. In February, they chartered a plane and flew back out to the Nation River.

This brings us to the dismantled airplane in the woods. Shortly before Brad and Lilly returned to the Nation, a bush pilot, dubbed by locals "George-the-Walking-Pilot," was flying along the Yukon toward Eagle and saw the packed airstrip on the Nation River. The runway in the snow was too much to resist. Besides, George was an old friend of Sage's. So he decided to drop in for a visit. The Piper Cub was on skis and George had no reservations about setting the light plane down on the river.

Hearing the airplane overhead, Dave and Sage went to their cabin door to see who it might be. As they stood watching, everything about George's approach looked good—until the last minute. The plane eased down, touched the snow and settled gently onto its skis. Then, without any inkling of a problem, the plane suddenly nosed into the snow and did a perfect forward flip, landing upside down at a forty-five degree angle. Dave and Sage, horrified and certain that George was hurt, bounded through the snow to the wreck, reaching it just as George crawled out from under. George was not happy.

Aside from the damage to both airplane and his feelings, George came out of the incident remarkably unscathed. However, he had two problems now, one obvious and the other not so. The obvious one was that his plane was upside down on a frozen river, which meant that when the ice went out in the spring, his plane would go with it. That might solve his second problem, the legal requirement that he remove his wreck from the river, but the ice would grind and chew and make scrap metal out of the airplane and George would prefer to remove it whole if he could. So the next day, the three of them laboriously pulled the plane back over onto its skis. With the tools George had on board, they removed the wing struts, the wings, and the engine, reducing the plane to a few manageable pieces. They hauled the pieces far enough into the woods to avoid spring flooding, and four months later we came upon them, wondering what the hell. Two days after his bad landing, George, a confident woodsman himself, put on his snowshoes and started the four-day trek to Eagle.

Where the Nation River widens its floodplain before meeting the Yukon, numerous small bog ponds and cut-off sloughs dot the wetlands. Buffleheads, mallards, and mergansers were abundant, and numerous tracks of moose, wolf, lynx, and fox riddled the muddy shores exposed by the low water. There were still bears in the neighborhood, too, in spite of the fact that a couple were shot earlier that summer. It was a productive landscape and both Eduardo and Garrett were kept busy with their species accounting.

Garrett occupied himself immediately with a small-mammal trap-census, something he did whenever we spent more than two nights at a location. Garrett was remarkable at his work: he could spot a vole runway through thick, matted grass or sedge; pick out miniscule scat from the duff of the forest floor; recognize shrew activity from the discarded wings and antennae of ground beetles. He could look at a shrub that had been browsed the previous winter and tell at a glance whether it had been eaten by a moose or snowshoe hare. He could break apart old wolf scat and, by identifying the hairs in it, determine what the wolf had last killed. He could do the same with owl pellets, naming its prey by the bones they contained. Garrett was an expert at catching animals and he had already made history on this expedition. While Mark and I were in Circle, leaving the others without river transportation, Garrett set some traps upriver from our base camp, and there, in a willow thicket on a low sand and gravel bar along the shores of the Yukon, he caught the first pigmy shrew ever to be recorded in a four-hundred-mile void between disjunct populations in western Alaska and the eastern Yukon Territory.

Piper Cub on skis, with wings and engine removed.

From the census work that Garrett was now doing, we would begin to piece together a rough picture of how energy and nutrients ebbed and flowed through the forest community of this area. The small rodents were the basis of much of the animal diversity here and as their numbers fluctuated, so too did the fortunes of weasels, shrews, foxes, and a number of raptors. Rodent populations in the subarctic forest are prone to cycle with some regularity as plant resources, most notably high energy fruits, also fluctuate. In years of high seed production, when plants have garnered enough carbohydrate reserve to support a substantial reproductive effort, rodents also prosper, obtaining sufficient nutrition themselves to reproduce abundantly, and often stimulated to do so by plant hormones ingested from seeds germinating under the snowcover.[6] So follow the predators until seed production declines again, rodent numbers fall off, and the predators suffer.

Garrett read into the animal sign he was finding, another story of predator-prey interaction: that of the snowshoe hare and lynx. It's a tale told in every ecology textbook, but much more has been added to this story of late. Hare cycles are usually explained as a classic case of predator-driven boom and bust: predators increase as prey becomes abundant, heavy predation drives the prey population down, predators then crash for lack of food, and prey recovers. But accumulating evidence suggests this may be only part of the story. The cyclic population fluctuations of snowshoe hares may be tied to changes in plant resources as well, and not just to the abundance of suitable browse. The idea distills down to this: Many of the plant species that hares depend upon during winter, particularly the dwarf birches and willows, produce mildly toxic or distasteful defense compounds in proportion to the amount of browsing the plant is experiencing. When hare numbers are down and browsing pressure is low, plants contain smaller concentrations of these defense compounds. Thus food quality is higher and hares produce more offspring. However, as hare numbers rise, browsing pressure increases and plants produce more defense compounds, causing hares to switch to plants of lower nutritional quality. This imposes a different kind of food shortage and by itself may slow population growth, but the turnaround at the peak of the hare population may still be driven by predators (predation stress by itself reduces reproductive success among the surviving hares), followed by a crash of the predator population. Once freed from heavy predation pressure, the recovery of the hare population may still be delayed, however, by high residual levels of defense compounds in plants, until the plants respond to the decreased browsing pressure after the crash of the hares.[7]

For the lynx, this added dimension to the story changes nothing. The lynx has become so narrowly specialized a predator that it has difficulty switching to other food sources when snowshoe hares become scarce, and thus lynx numbers parallel closely the population cycles of their principal prey species. According to local trappers, the snowshoe hare population on this stretch of the Yukon crashed during the winter of 1973–1974, and the following winter lynx were so hungry they would come to any bait.

While Garrett was piecing together a clearer picture of the present, Bruce was busy digging up the past, trying to understand where this forest came from; and in the layered turbidite deposits of the Nation River formation, Bruce found something better than gold. One afternoon, when I had gone a short way down the Yukon with him to assist at a shoreless outcropping on the river's edge, he chipped his way through a thin layer of shale into the storage vaults of the long lost Devonian Period, the golden age of plant evolution some 350 million years ago. It was a fascinating discovery. As I steadied the boat, Bruce chinked away with his hammer, knocking bits of fossil-bearing rock out of a stratum near the water's edge, interpreting aloud the ancient sedimentary environment of the formation as he examined each fragment. I found myself back in the Paleozoic Era on the beach of a warm-water sea that lapped at the base of steep slopes. The rains were more generous then and had washed much sediment and plant material into the receptive waves. The larger stems, branches, and trunks sank quickly and broke up easily in the surf, and were soon incorporated into the coarse shoreline sediments. Smaller plant material floated further out where it eventually waterlogged and settled into thick layers of finer sediments, leaving its impression in today's mudstones. In a few places, peat deposits had formed along the shore and as the waves wore away at their underpinnings, large chunks of organic material collapsed into the water, eventually becoming buried and preserved as bodies of coal. I could smell the salt air, hear the waves lapping, feel the heat.

All that came to pass more than three hundred million years ago, and now the river was digging back into the archives. The river, it seemed, was a force of uncompromising persuasion, forever carrying the past into the future. There was no present, save for a sand grain coming to rest for an instant before being struck by another and sent on its way into the next layer of earth history to be squeezed up from the ocean floor. From source to sea and back again, the past was continually being taken apart and reassembled. As we hung to the side of the outcrop that afternoon, Bruce worked to intercept a bit of that past, that we might better understand the ground we slept upon.

Back at camp that evening I was sitting in the sand, leaning against a log and still musing over the Devonian seascape when Bruce snapped me out of it. Bursting through his tent flap, specimen in hand, he dropped a dark rock on the ground at my side.

"*Protolepidodendron.* Look here." he said, squatting down to show me. "See these decurrent leaf impressions? Their bases are enlarged to form an elongated, obovate cushion and they're closely packed in a pseudowhorled spiral. Definitely *Lycophyta.*"

"Whoa." I said, grinning at him. "You wanna run that by me again, in English?" I couldn't help poke a little fun at him, but it was with respect and he knew it.

"It looks a lot like that staghorn club moss," he said, gesturing toward the ground under a nearby spruce, where a carpet of light-green, trailing plants softened the forest floor. He stood there for a minute as if seeing something the rest of us couldn't. With distant thoughts he turned and walked slowly back to his tent.

I stretched out on the ground and stared off at the spruce along the river, unconsciously rubbing the grainless mudstone that Bruce had tossed at me. I squinted until the trees lost their detail and merged into the ghostly canopy of a tropical forest on an ancient shoreline. The club mosses under the spruce, so much like their earliest ancestors, trailed off into the obscure haze of my vision to form a dense ground cover along with tall horsetails that reached upward into the understory of the primeval forest. Another *Lycopsid*, not much different in form from the low club mosses, grew to giant proportions under the forest canopy, reaching twenty or more feet in height, leaving me feeling like a gnome under its drooping foliage. Towering over these tree-like club mosses were majestic conifers, prototypes to rival any of their descendants, growing in dense clumps with trunks over three feet in diameter and spreading crowns of huge fern-like leaves. At their base, in the organic duff of the damp forest floor, fungi were quietly at work developing their own specialty, breaking apart old blueprints and freeing molecules to try again at something new. It was an exciting time in the evolution of land communities.

If this part of Alaska was tropical three hundred fifty million years ago, it seemed scarcely less so now. The air hung heavily during our days on the Nation, the temperature rising into the mid-nineties and the humidity with it. The interior basin of Alaska is topographically bounded by the Brooks Range to the north and the Alaska Range to the south, and consequently experiences a notoriously continental climate, which means hot summers as well as frigid winters. Periods of prolonged calm reign year round as weather systems from the Arctic Ocean and North Pacific are blocked. During summer, air masses lifted over the Alaska Range drop much of their moisture on the coastal flanks of the mountains and then warm as they descend the northern slopes into the Interior. It was a double curse, for we had to stay fully covered much of the time against the omnipresent mosquitoes, but each day the air temperature pushed a little higher until finally our sheltered thermograph at ground level broke one hundred degrees. And with the increasing humidity, afternoon clouds began to build. A few thunderheads developed here and there, with rumblings occasionally, but it wasn't quite their time yet.

In heat like this, the presence of permafrost anywhere in that country seemed incongruous, yet one afternoon when I was poking around in a nearby bog, I found frozen soil only twelve inches below the surface. Disbelieving, I went back to camp for my thermistor probe and returned quickly to see if I could verify my find with a few soil temperature profiles. Sure enough, in this hundred degree weather, the soil beneath the moss layer was frozen.

For the next two days I sought relief from the heat in the bog that had just captured my fascination, immersing myself in the most literal way with my data collection. Sloshing around in the tea-colored water with a topographic level and a makeshift surveyor's staff—a spruce pole with centimeter marks notched into the bark—Mark and I mapped the contours of the bog, its hummocks and depressions, and the depth to frozen soil. Out in the center of the bog was a sizable area of open water, waist deep, and we probed it, too, in order to determine its contours. Sinking my temperature sensors as deeply into the bottom muck as my leads would allow, I attempted to use temperature data to extrapolate depth to the permafrost table, and I found an unfrozen window in the hard basement. It was a textbook study in the workings of the boreal forest.

The bog had apparently formed in an old oxbow pond that had been cut off from a meandering channel, probably during spring floods. There would not likely have been frozen ground beneath the bog in the beginning, due to the significant heat storage capacity of the standing water. Things began to change, however, with the encroachment of aquatic vegetation around the margins of the new pond, especially with the long, intertwined stems of pondweed and buckbean reaching out over the open water. Such encroachment would have occurred slowly at first, probably not affecting the unfrozen ground for a while. However, as organic sediment accumulated in the shallower areas, it provided firmer ground for new arrivals and a procession of plants followed, one species or group of species replacing another in time as the physical nature of the pond edge gradually changed. Slowly, the pioneering aquatic species were displaced outwardly toward the center of the open water by colonists with more terrestrial affinities, until eventually the site became suitable for invasion by one or more species of sphagnum moss. This is the classic process of plant succession in northern bogs.

Once sphagnum, or peat moss, came on the scene, things began to change more rapidly. It was common knowledge among indigenous people and white settlers alike that peat moss, when it is dry, has superior insulating properties, and was the chosen material for chinking and insulating log cabins. As peat moss grew over the vegetated portions of the bog and dried out during the summer, it insulated the ground too, preventing the summer heat from penetrating into the organic soil beneath.

But peat moss has a dual personality, behaving very differently when it is wet. Peat has extraordinary moisture retention capacity, able to hold about ten times its dry weight in water, which is why Athabaskans and others across the boreal forest packed it around their infants in the cradle board. It was a natural diaper. When wet, however, peat moss becomes a very efficient conductor of heat due to the physical properties of the water it contains. Thus in the fall, when evaporation rates are lower, the peat in the bog remains saturated, later freezing, and all winter long it facilitates the loss of heat from deep within the ground. The ground freezes hard, and the following summer it is insulated again by the drier peat, slowing its thawing. Gradually, as the bog vegetation closes over the open water from the sides, the deeper permafrost will also encroach toward the center.

The story might end there except that in a strange turn of events, what starts out as a topographic depression can end up a huge earth mound rising thirty feet above the surrounding muskeg. In the final stages of bog formation, when the pond basin is filled with sediment and sphagnum covers the entire surface, frozen ground will begin to persist beneath the moss, gradually extending downward. With permafrost encroaching now from all sides, groundwater becomes trapped in the unfrozen core beneath the former pond. Confined within the closing sphere of impermeable permafrost, the trapped water even-

Bog pond gradually filling in with the encroachment of aquatic plants and sphagnum moss.

tually comes under great pressure, and when it finally freezes, may exert enough force with its expansion to rupture the surface, bulging the ground upward into a mound known as a pingo. The exposed dark soil may then thaw as it absorbs sunlight, collapsing the center of the pingo and impounding water. Aquatic plants invade and the cycle turns again. Numerous pingos existed at the headwaters of the Nation River and Fourth of July Creek.

For most of our stay on the Nation River, I was preoccupied with a stand of white spruce growing on a steep hillside just beyond the cabin on the right bank of the Yukon. Like Kathul Mountain, this south facing exposure was dry and lacked any discernable permafrost. Here, too, sage and large patches of common juniper covered much of the drier ground. But in the crease of a shallow drainage grew some of the largest white spruce I had yet seen; not trees of Pacific Northwest proportions, but spruce a hundred feet tall and twenty inches in diameter. It was their age distribution and growth rates that I was mostly interested in, as a yardstick against which I might evaluate growth of the treeline stands that I would soon be studying. Treeline in this area was interesting to me in that it appeared to be advancing up the mountainsides as if not yet fully adjusted to the present climate of interior Alaska. If this turned out to be the case, it might have implications regarding post-Pleistocene vegetation changes and the present range distributions of some boreal plant and animal species. But to test some of my ideas would require knowledge of what the growth and reproductive potential for white spruce was under optimum conditions, and from what I had observed so far, this seemed like a best-case scenario. So I established a series of sample plots along the entire gradient, from river's edge to the top—a fifteen-hundred foot rise along what seemed at times like a vertical hillside—and every day, Mark and I would climb the steep slope, drenching in sweat, to measure and core the trees in my plots.

The wind changed direction our last evening in camp at the Nation River, and on it was the faint hint of fire. It wasn't quite the smell of wood smoke yet; it was much lighter, a thin, yet unmistakable haze in the air. In the subarctic dusk it created a paradoxical serenity along the river. Like vapor rising from the tepid Blue Ridge Mountains after a rainfall, the haze softened the distant trees, silhouetting those nearest the riverbank so that the forest took on an ethereal quality, a simpler beauty. But somewhere in the distant and endless spread of boreal forest, a fire was racing through the muskeg as crown after crown flared in yellow-hot rage and then died down. The fire created its own winds, fanned by its temper, and roared madly on, not even taking the time to finish its doing. Branches were stripped bare of foliage, but even fine twigs were left unconsumed, especially near the tops of the trees where cones were densely clustered. It was enough, though—it was what the black spruce were waiting for, the heat of fire opening their cones to release the seeds of the next generation. Then, in a momentary cloudburst, the fire somewhere on our blind side was extinguished, ended by the same force with which it began, as if correcting a mistake. The season of thunderstorms and wildfires had begun.

V. CHARLEY RIVER

We were all a little impatient, tired of waiting and anxious to get on with the airlift. We had broken camp and packed up all our gear right after breakfast, and spent most of the day sitting around, listening, and watching the skies. By early evening I was sure that something had gone wrong.

It had been three weeks since my last communication with Gordon, our helicopter pilot. We had met in Fairbanks to lay things out as thoroughly as possible, going over every detail of our move: what day we wanted to be picked up, how he would find us on the Yukon, where we would establish our next base camp in the mountains, how much gear we would have. The plan was that Bruce would stay on the river with the boat and would be joined by a geologist from the University of Alaska for additional collecting on Seventymile River. The rest of us would move to higher country and work the alpine area around Twin Mountain. With no way of communicating once I left Fairbanks, it was imperative that we stayed within our plan. Keeping track of time became an important task in the perpetual light of the subarctic summer, as we often worked right through the night. so we carefully marked our calendar while on the river and today was the appointed day for our air support.

Gordon was young for a bush pilot, but he had gained more than enough experience flying in Vietnam and had earned a reputation for being reliable. His helicopter was a small, conventional-fuel machine, capable of carrying two passengers plus the pilot and a limited amount of gear. We had talked things over carefully and had decided that with two shuttles from our river base, we could get ourselves and the equipment we needed into the Twin Mountain area without difficulty. After the first drop-off, Gordon would detour back to Circle to refuel, then return upriver for the second move. It sounded workable and I had confidence in Gordon.

Just about the time I decided he wasn't coming, the first faint throb of a helicopter engine filtered through the quiet noise of rippling water. All at once we were on our feet. We waited without a word, watching the horizon. As the sound slowly grew louder, our anxiety heightened. There was an element of excitement in our pending move; each of us felt it, and the long wait seemed only to amplify it now. When the helicopter suddenly appeared above the skyline I could see immediately that it was not Gordon. It was, nonetheless, coming for us.

The big machine circled once and came down slowly onto an open stretch of the bar about fifty yards from where we stood, whipping the sand into a stinging, choking whirlwind. It was an outrageous violation of our peace. As the chopper blades whined down, the pilot hunched out of the cockpit and stooped briskly from under the threat of the whirling knives. I waited for him to speak first.

"You Peter?" he asked.

"Yes."

"Name's Mike. Gordon sent me. He's havin' trouble with his starter and couldn't make it." He lit a cigarette.

I extended my hand. "Glad you found us, Mike. Gordon tell you what we want to do?"

"Sure. Couple of you want to move up the Charley, huh?"

I faltered for half a second. "Four of us and the dog. And there's our gear," I said, waving to the stack of boxes and duffels.

He stopped short and took the cigarette out of his mouth. "Four of you and all that?" he said with an unsettling note of surprise.

"We had planned to do it in two moves."

"I haven't got enough fuel to do it in two moves," he answered.

"What about Circle? Can't you re-fuel there?"

"This is an Alouette turbine," he said, jerking a thumb in the direction of the helicopter. "I need jet fuel."

"There's no jet fuel in Circle?"

"No. Besides, I'm due back in Fairbanks in four hours."

I was exasperated. The others were looking at me and I didn't know what to say. Mike obviously felt it, too, as he walked over to our pile of gear and looked it over. He poked at a box with his foot and hefted a duffel bag. We waited for a comment.

"Well," he said hesitatingly, still calculating. "We'll be heavy, but I think we can make it."

I should have said no right then. This guy was too cocky behind his mirrored glasses and cigarette, and I didn't like the situation a bit. But I had to concede to his judgment. I didn't have any experience by which I could challenge him. Neither did the others. They just looked at me because it was my decision, and all I could do was shrug.

"Okay," I said. "If you're sure."

With that consent, we began hauling our gear over to the helicopter where Mike watched us load it, piece after piece, onto the cargo racks. We knew it would be a tight squeeze in the cockpit, so we tied everything on the outside—rifles, sleeping bags, food—and when the last piece went on Mike remarked cheerfully, as if to offer the only encouragement he could, "I didn't see you guys carrying anything too heavy." An absurd comment, I thought at the moment.

With everything secured, we shook hands with Bruce and climbed into the helicopter: Garrett, Mark, and I in the back, and Eduardo, with the dog at his feet, next to the pilot. Mike gave a quick check around, slipped his padded earphones on for protection, and started the rotors.

It was immediately apparent just how heavy we were. Mike wound the engine to a painful pitch, then strained hard to lift the helicopter a foot or two off the ground so that he could move it up to the far end of the river bar. Too heavy for a vertical takeoff, he was maneuvering to attempt a run at it like a fixed-wing airplane: gaining speed, momentum, and then lift. When he had the running room he wanted, he turned the machine around, revved the engine ever higher and started down the bar.

It was slow at first but then we picked up speed, faster, faster, until the sand under us was a blur. The willows flashed past. We were hurtling toward the river, but for all the awesome power and frightening roar of the jet engine, still the willows were at eye level. We were not gaining any lift and running out of room. Coming up fast at the end of the bar was an ominous tangle of logs, whole trees ripped up from the river banks in the spring floods and left stranded there, and we did not have the elevation to clear them. Mike pulled back hard and set down right in front of them.

I don't know why I couldn't find the words to say, "Stop. This isn't going to work." Instead, I let him try again. We turned around and taxied back up the bar for another attempt. The noise, the heat, the stress inside the cockpit was almost unbearable. Mark was on the verge of breakdown. He had his face buried in his huge hands, groaning, and I was clutching his arm hard, trying to keep him together, trying to console him. "Hang in there Mark, hang in there." I kept saying, and all the while I was trying to rationalize this insanity in my own mind while we were whipping up a dust storm outside. The engine screamed and Mike showed no mercy. The man seemed desperate to succeed as he asked for absolutely everything that machine had to give. This time he managed to gain a few feet of lift before starting down the bar, and one more time we threw ourselves at the log jam.

We watched the driftwood pass under us as we barely cleared the snags. We were airborne, just skimming the willows and then the tops of the spruce lining the Kandik. Out over the open muskeg beyond, we found a little more flying room but stayed on the right bank of the Yukon for quite some distance before Mike felt he had enough elevation to cross the big river safely. I wondered how he thought we would make the summit of Twin Mountain, but maybe he knew what he was doing.

We followed the lower Charley for a while as it meandered through the lowlands to meet the Yukon, intending then to turn up Hanna Creek, which would lead us directly toward the summit of Twin Mountain. My attention quickly shifted to the terrain below. From the air, the complicated mosaic of vegetation patterns that we had been seeing along the river made immediate sense. In their constantly shifting courses, the rivers cutting and eroding into their banks on the outside of their bends would simultaneously slip off the inside of the bend, leaving fresh deposits for plant colonization. Every river bend had associated with it a history of shifting courses revealed from the air by a series of concentric arcs paralleling the river and representing vegetation of different ages or stages of development. It was a fascinating landscape, though my enjoyment of it was short lived.

Thunderstorms were breaking out around us. The peaks were already lost in the clouds, and we could see that we were going to fly into some heavy weather up on the flanks of the mountain. Sightseeing over, the tenseness of our situation returned. With rain lashing at our cockpit, we concentrated on the drainages, following them on the topo map, counting. Bonanza coming in on the left, then Cultas on the right, Silvia, Ericson, Dewey; slowly we checked them off as best we could see them. Then came Hanna Creek, and we turned toward the mountain.

The visibility was considerably poorer as we started up that rocky drainage. The clouds boiled in towering, turbulent drafts. Twin Mountain was only five-thousand-seven-hundred feet in elevation, but that was about three-thousand feet above our present level and it looked like an impossible gain for our load in this weather. Still, Mike stayed on course.

I couldn't see very well straight ahead, but suddenly there was a commotion up front and Eduardo shouted to Mike, "TURN AROUND!" Immediately Mike yelled back, "I CAN'T. HANG ON!"

For an instant I caught a glimpse of the mountainside ahead. We were closing in on it fast, and Mike, unable to gain elevation, had waited too long to concede. He had gotten himself boxed into a steep-sided ravine and couldn't pull out. Eduardo looked down through the bubble cockpit. Twenty feet below there was nothing but rock and water.

In a remarkable display of skill, Mike brought that helicopter to a standstill in the air, the big rotor pinging as it clipped the brush on both sides of the ravine. Slowly then he eased us downward until one skid touched a huge boulder in the streambed. As if the skid were an iron hand he clung to that boulder for a modicum of stability, and then came the order. Mike's only escape now was straight up and he had to lose weight fast. Above the deafening throb of the blades he turned back and shouted "JUMP OUT!" I couldn't believe it.

There wasn't any time to discuss a new plan. Eduardo banged the door open and shoved the dog out, and though I hardly remember doing it, three of us dropped out of that helicopter into the creek below. It was dizzying—the lightning, the rain, the deafening jet engine and furious wash of the rotor as the helicopter lifted out of there—and the three of us stood in the water looking straight up as Garrett and everything we owned disappeared into the clouds.

We waited on that creek bed, unhurt but stunned with disbelief, each of us struggling to understand what was happening. In perhaps forty-five minutes, maybe a little more, we heard Mike coming back. In another minute he appeared from downstream, flying low, but staying out of the ravine. His cargo racks were empty.

I didn't see any way he could pluck us out of there and I guess Mike didn't either, for he just hovered over us for a few seconds, maybe two hundred feet above, and then turned back downstream. In a couple of minutes he reappeared from the same direction and hovered over us again. If he were trying to signal us, we could not see him, but when he turned back downstream a second time, we got the message. Garrett must be down on the Charley. It would have made perfect sense for Mike to go downstream after dropping us, rather than attempt the summit in thunderstorms. He could return tomorrow and try again in better weather. When he appeared for the third time, Mike was considerably higher. He paused over us for a brief moment and then was gone, this time we knew for home.

When we had boarded the helicopter back at the Kandik, a few loose, last-minute items were thrown in the cockpit with us—camera, my briefcase, a box of almost forgotten cooking utensils—and when we bailed, they tumbled out with us. So we gathered up the odd items and with unspoken consensus started walking downstream. Because the terrain was rough and the brush thick, we just stayed in the creek bed. It must have been a sight, the three of us with a dog and little else, a hundred miles from anywhere, soaked and slogging through the water in the grey din of the storm. But I was inexplicably content at that moment, leading the procession down the mountainside, briefcase in hand; relieved, perhaps, that we were alive and out of the helicopter. After a spell, Eduardo broke the silence.

"For Christ's sake Marchand, you look like you're going to the office."

I laughed. It was a long way to the office, though. We figured we were about three miles above the Charley.

As I turned and threw a grin back toward Eduardo, I noticed that Mark was struggling. He was in bad shape emotionally and lagging seriously, and we needed to get down quickly. The helicopter ordeal in itself was terrifying to Mark, but I sensed that the uncertainties we faced now were even more troubling. In spite of his experience—Mark had trained back East for just this sort of wilderness exigency—he was not holding up well at all. I hoped more for Mark's sake now that Garrett would be waiting for us with a pot of soup on.

It was near 10 pm when we finally reached the Charley. In the last hour, we descended several hundred feet in elevation into a steep-walled gorge while the storm came down hard, lightning striking the ridge above us with dazzling flashes and bone-rattling thunder. We had long been drenched and now the wind was picking up. A twinge of fright came over me when I didn't see Garrett immediately.

Between the wind and the rush of water in the Charley River canyon, it was hard to hear anything else. We hollered loudly. When we didn't get a response, I pulled a whistle out of my pocket that I carried to alert bears when I was working alone, and blew it with a renewed sense of urgency. There was no sign of Garrett. We looked quickly for the cabin; according to our sources, there should have been a trapper's cabin at the mouth of Hanna Creek. But there was no cabin either. It was beginning to sink in that we were in trouble. All three of us were shaking. Whether our emotions had just kicked in or we were on the brink of hypothermia didn't much matter. We were wet and cold, we had to get warm quickly, and exercise was no longer sufficient. We needed a fire and everything was soaked.

In my briefcase I had a paperback book. I fumbled for it, clumsily spilling much of the case's contents while trying to keep the book out of the rain. Kneeling, hunched over to shield it, I paused for a moment at the title—Le Carré's *The Spy Who Came in From the Cold*—then started tearing and crumpling the pages, stuffing them between my knees to hold them. With a sense of urgency Mark and Eduardo gathered bunches of small twigs and whatever other burnable fuel they could find from the undersides of deadfalls. My cold hands shaking with tenseness, I moved a rock aside to expose drier ground beneath it, set the paper down, and laid the tinder as carefully as I could in a loose pile, poised over it like a hooded cobra.

"Okay," I said in a deceptively controlled tone. "Matches."

"Matches. DAMN!" Eduardo cursed. I glanced up to see a horrified look of disbelief. "MY MATCHES ARE GODDAMNED SOAKING WET!"

"Wait a minute." I straightened a little and shoved my hand into my tight, wet pocket. Only a couple of days earlier my waterproof match case was in my rucksack in the boat when it occurred to me that if we had a spill and my pack went floating downriver, I'd be in a mess without my matches. So right then I groped around for them and stuck them in my jeans. Two days later I was still wearing those jeans. They were all the matches we had.

I lit the small heap of dry paper and damp twigs, crowding it to keep the drips off, and nursed it piously. "C'mon, c'mon," I muttered in fervent prayer. Painstakingly, we added thin shreds of damp birch bark and bits of lichen, then slightly larger twigs and more bark, and the flame licked at the damp fuel. We started breaking deadwood from lower branches of spruce trees. We hauled driftwood from the river's edge, bigger and bigger. The fire took and I breathed again. We built it into a bonfire and stood close for a long time, backs to it, then facing in. The rain let up and we stripped, drying our clothes over the fire and stocking up on its warmth. We felt safe for the moment, but it was a troubled respite.

A sharp hunger pang brought renewed focus to our situation. Back at the Kandik River, we had packed up our food and cook gear early in anticipation of the move. We'd left a bit of food out to munch on during the day, but were holding out for a hefty meal when we set up our new camp in the high country. It was already late in the day when Mike showed up and now several more hours had elapsed since we left the Yukon. Full of anticipation hiking down the stream, our thoughts of finding Garrett with the soup pot on were quickly superseded by the exigencies of our situation. Warm now, and dry for the moment, the hunger came back with painful sharpness, no doubt amplified by reasoned fears that Garrett was nowhere around. We had control of our situation for the moment, but our comfort was shallow; and though

Drying clothes after bailing from a helicopter in a thunderstorm.

food was perhaps not our greatest physical need, it would have provided a welcome sense of security.

As soon as we felt sufficiently relaxed, Eduardo and I decided to split up and cover the river in opposite directions looking for Garrett, leaving Mark to tend the fire. We agreed to walk half-an-hour out and then turn around. An hour later we were back; no Garrett and no gear. We stood by the fire for a while longer, going over and over in our minds the sequence of events that led to our predicament. Where could Mike have dropped Garrett? What else could he possibly have been trying to convey when he hovered over us and repeatedly turned downstream? Why isn't Garrett using the rifle to tell us where he is? What does he think we are doing right now?

From the high and windy bench on Twin Mountain where he had been dropped, Garrett caught intermittent glimpses of the helicopter well below him. He watched Mike through the roiling clouds make a pass over the drainage where Garrett thought we were. That Mike's first two approaches were from the direction of the Charley didn't register anything with Garrett. Mike had expressed doubt that he could get back down into that ravine, but said he would motion the three others to hike upstream to where Garrett was.

It was 7:00 pm, still plenty enough light, Garrett thought to himself, even with the heavy clouds. If he were to hike down the south ridge from where he stood, he could intersect the upper end of the creek and follow it until he met us. He figured we could use a little protection from the weather, so he hunted up our rain jackets, then stacked all the camp gear into two piles and covered them securely with tarps and a tent. Stuffing a little extra food into his daypack, he picked up the .444, checked the magazine, grabbed a few more shells, and set out for the creek. He reached it with half-an-hour's bushwhacking through the thick and wet shrub tundra and started down.

For another hour, Garrett descended the creek bed until he was certain he was past the point where we had bailed out of the helicopter. He was miffed that he had seen no trace of our presence. Garrett was a disciplined observer; his reputation as a field mammalogist had much to do with his ability to read the subtlest of animal signs; and humans, by comparison, usually left blatant evidence of their passing. He would have noticed a kicked up rock, a water-filled boot impression in the moss, scuffed spruce needles, disturbance to the smooth gravel in the stream bars. But if we had left any hint in the creek bed of our destination that night, the rains by then had erased all traces. Garrett turned around and started the difficult climb back to his new campsite. Hours later, exhausted, he reached the bench, and his last hope that we might be waiting for him dissipated. Five times during his fruitless trek he had fired the .444; his last act before setting up a tent and crumpling wearily into his sleeping bag was to place a lantern on top of a lone boulder looming prominently over the high tundra. Its bright little flame was a lost beacon in the night. Garrett was alone now, and he knew the helicopter would not be coming back.

After a few hours Garrett awoke anxiously. His first thought was to try again to find us, so he slipped into his wet boots, grabbed the rifle and pack, and started back down the ridge. But the creek was flooding its banks with the storm water and it was no longer possible to descend in the channel or to cross it safely. Feeling lost himself and still bone-weary from physical and mental effort, Garrett returned to camp. There, the frustration and the questions set in. He took out his journal and began writing, as if by methodically laying out the details he might see a solution. And in the dialogue with those pages, his mind began reeling.

"Why haven't they come up to the high country and seen my signal? There are three of them. Even if one is hurt, one of the others could come up looking for me. Where are they?" Then he slowed himself down.

"Okay. What do they know?" He tried to think through the last minutes in the helicopter. "They know that I was dropped off with the gear on the South ridge up from the stream. They know I am alone and can't rescue all of them." He stopped and studied his watch.

"It's been eighteen hours since we separated. Where are they?" he repeated himself. Then he looked over at the one stack of supplies still under the tarp.

"I'm safe enough," he wrote. "I've got food, shelter and warm clothing." Pause. "Mosquitoes are bad all over. Damn, I don't know if they even have bug dope. They'll have a hard time surviving with nothing but light clothes."

He sat for a while, then pulled the axe out and walked down to a clump of isolated, dead spruce. Compulsively, he chopped wood and brought it up to the bench, shoving it under an overhang of the rock that had become his signal tower, trying to keep the wood out of the weather. All the while he kept an eye on a thunderstorm moving toward him, but it still took him by surprise. When the rain started he grabbed the lantern and the axe and bolted for the tent, but he was soaked by the time he got the zipper open and dove in. He'd left the storm flap open on the rear window, too, and already his sleeping bag was getting wet. "I've got to be more careful," he thought. He picked up his journal and started writing again, as if talking to an unseen partner:

"I'm getting afraid for the others now. What the hell are they doing? Where are they? My God this is terrible."

We kept the fire burning vigorously all night, its flames seeming somehow to keep the worst of the rain away from us through the intermittent showers that continued. Mark slept on the ground soundly,

Eduardo and I cursing his ability while every now and again jumping to extinguish a spark on his clothing. It was a fitful night for us as we retreated for a while into our own thoughts and tried to nap. As the still-turbulent sky began to lighten with the morning twilight, we talked again about our situation. We decided to hike farther down the Charley, Eduardo and I together. Maybe we weren't on Hanna Creek, we thought. Maybe the cabin is below us and if we could get to it we might at least find some food in the cache. We left Mark again with the fire and were gone quite a while this time, perhaps seven or eight hours, and figured we had covered four miles or better downriver. Still no Garrett, and still no cabin.

An undercurrent of fear began to permeate our thoughts. If Mike were coming back to correct this situation, he would have done so by now. If Mike didn't make it back to home base in the heavy storms with his damaged rotor, the FAA folks would be all over the place looking for him. And when they found him, maybe he'd be able to talk or maybe not. If he did make it to Fairbanks but didn't see fit to send someone else back out here, we could only assume that he reported everything was A-ok, damage to his machine notwithstanding, in which case no one, not even Mike, would know where we were. That leaves Gordon or someone else to find us, but not for a while yet, and that assuming our request via Mike to be re-supplied in five days was somehow communicated to Gordon. Playing guessing games with limited data is not an unfamiliar role for field scientists, but playing guessing games in a deep, cold, wind tunnel carved by a wild river in the middle of three million roadless acres, while shivering and hungry, was not something we were prepared for.

The day passed. We speculated endlessly, busied ourselves gathering wood, huddled around the fire again during heavy afternoon showers, and talked about tomorrow. Mark was becoming despondent and there wasn't much we could do to change his mood. He slept a good part of the time and talked little when he was awake. When the next day showed no more promise than the previous, we moved our campsite to a broad bench beside a wide, cobble-paved bend of the river, dry this time of year, to give ourselves more visibility, and then occupied ourselves building a shelter to mitigate the now daily thunderstorms. There was no shortage of dead tree saplings in the dense spruce stand

Keeping dry on the Charley River, with no food or gear.

along the river; mere poles devoid of branches and bark, too crowded to fall over even long after they succumbed to competition from taller neighbors. Nor were we lacking an ample supply of spruce roots—long, cord-like, pliable roots strong enough for lashing—lying just below the surface of the loose, damp organic soil. So we rounded up poles, dug roots with our hands, and fashioned a lattice that we leaned against a stout crosspiece tied between two trees. With my Buck knife, the one tool at our disposal which I'd kept permanently affixed to my belt, we peeled broad sheets of outer bark from several large birch in the area and shingled them atop the lean-to frame, weighting them down with brush and more poles. Three larger bark sheets made good ground pads; uninsulating, but at least waterproof.

The reward was immediate. The new shelter kept us dry that afternoon and gave us a much needed, if somewhat inflated, sense of control over our situation. Building a small fire just outside the lean-to also drew some of our attention away from our discomfort, and helped with the mosquitoes. Still, it was too cold at night for Eduardo and me to sleep, even next to the larger fire that we kept burning continuously. So the two of us would talk through the darker hours, watching Mark and the fire, until sunlight spilled over the eastern river bluff around five in the morning. The night's chill gave in quickly to the sun and sleep came mercifully.

Our most realistic hope now was that we would have to wait it out only five days on the Charley: that somehow, when Gordon returned to the Twin Mountain area with more supplies and didn't find anyone there, he might start looking around. That was assuming he got the message from Mike in the first place. Our best chance to get out sooner would be to get lucky and signal a passing bush plane. And that's assuming someone is out there flying around in this part of the country. So every day we diligently moved our fire onto the wide, slip-off bend of the Charley, where we would be in full view from the air, and waited on the rocks with piles of green willow and alder brush.

On the third day we got a bite. Sometime after noon, we thought we heard the drone of an airplane engine above the rushing water. In unison we froze on the rocks, straining for more information. It seemed to be getting louder. Then we spotted it: a small glint of silver in the sunlight, heading more or less in our direction. We jumped up and piled the brush onto our fire. Smoke billowed upward. The plane flew past, high, and disappeared over the horizon, but just before it faded out of sight, we thought we saw it bank to the left. Could it possibly be coming back? We waited, arms full of brush, watching the treetops.

Suddenly the plane was there, coming right down the chute in front of us. It roared overhead, smoked billowed again, the three of us waved and hollered like madmen. We had reeled him in. The pilot would be sending out a radio call right now and someone would be looking for us in a matter of hours.

By midnight, discouragement had settled in like the cold, damp air filling the valley. We couldn't understand why no one had shown up. We sat morosely around the fire, long spells of private thought punctuated only occasionally with talk.

"Damn, if I ever get out of here, I'm gonna go straight back to Seattle and ask Pam to marry me."

"I've been thinking about buying a piece of land and building a small house when I get back to Vermont, something where the kids'd have room to tromp around in the woods—and maybe a greenhouse for Janet."

"Boy, you guys gotta spend a night in Seattle when we get out. We'll go down to the fish market, get a bunch of crabs, mussels, scallops, maybe some halibut, and cook ourselves up one hell of a feast."

"You got a deal there man, but a big ol' beaver'd suit me just fine right now."

The talk always seemed to turn to food.

And always in the back of my mind was Garrett. Where could he possibly be? What did he and Mike talk about when he was dropped off with all the gear? We couldn't guess any more. We weren't even sure where we were. I looked over at Mark, his back against a tree. Big, sullen, unresponsive; he was holding his left wrist, thumb pressing down, checking his pulse again.

Breaking spruce twigs and tossing them absent-mindedly into the fire, I spoke up.

"I know one way to get some attention out here."

"How's that?"

"Torch this place. BLM'll be on it in no time"

Silence.

"I guess." Long pause. "I don't know. Maybe we're not that desperate yet."

"Thanks. I was hoping you'd say that. You'd have to light the match, 'cause I couldn't do it."

"Let's just wait it out a couple more days and see if anyone comes looking for us. If they don't we'll figure things out then, but staying here seems like the only chance we've got of being picked up."

"Yeh."

Silence.

I got up slowly and walked out on the rocks to the river's edge, hunched down on one knee as if to take a drink, and I could no longer push it back. In the reflection of clouds off the unsympathetic water I saw the faces of Janet, Greg, and Danielle. I missed them, wondered if they'd received my letters. A tear broke loose; a choked sob; then another. The rushing Charley absorbed my tears, carried them to the Yukon, drowned the sound of heartache.

Garrett emerged from his tent at 6 am. He went to the rock and put out the lantern. From his position above treeline, Garrett presided over a vast uninhabited landscape. His view to the east was unobstructed for a distance of fifty or more miles. He pondered now the expansive range of mountains extending clear to the Yukon Territories; disquietingly empty at the moment, yet a strangely tranquil scene, a coarsely textured mosaic of spruce forest and tundra, colors shading from light green in the foreground to blue-gray in the distance. In the middle ground, from Garrett's vantage point, was the deeply incised canyon of the Charley, one of the larger tributaries of the Yukon River. Following its contours to the northeast, he could see two bends several miles distant, where light mirrored off the water. Just this side of the first bend was the crease where Hanna Creek hid itself, and below him was the unnamed drainage that the five of us had attempted to navigate in an overloaded helicopter two days earlier. The distance was deceptive, the silence deep. Now he was very much alone.

Garrett fired a single shot from his .444 and waited for a few minutes. Then he went back to his journal. "No sign of anybody, anywhere. All clouded over. A bad day." He fixed himself some breakfast.

After cleaning up and securing everything as best he could against the weather, Garrett loaded a pack, took up the rifle, and set out again as he had the day before. Returning to camp many hours later, he slumped down wearily and talked to his journal.

"Left camp at 9:00. Walked along ridge and then cut down toward the creek. Got to the creek and followed it—almost reached Charley River. No tracks or sign of the 3 men walking along creek. If they were following it down this side there are some obvious ways to go and I think I'd see traces of them."

"I went back up creek sticking close to it, [illegible] right at edge and always in sight about 20–30 yards back. No sign. If the Labrador was around he'd detect me. When I got about half way up, where I've been the last 2 times, I cut up midslope. No sign of them at all. Returned alone to tent about 3 pm—tired out. Now I don't know what else I can do. The 2 tarps hanging on the big rock up here were visible from all along the ridge and from part of the creek bed. I can't figure out why they haven't seen them or tried to get up on this ridge to [illegible]. At least I know they are not lying hurt along the creek. I've walked along every bit of it—two and three times for the center section—not a trace of them. I feel guilty not being able to find them, but I don't know what more to do now."

The next day Garrett walked down to treeline, cut three large spruce trees, limbed them and dragged them up to his camp. There he peeled the bark off one side to expose the glistening white sapwood and laid the trees in a straight row, spaced several feet apart and parallel to one another, as a signal for help. Then he almost got lucky.

"A plane flew down the Charley River. I heard it about 2 pm and dashed out of the tent to look. A silver-colored, single-engine plane flew right past the mouth of our creek and down past mouth of Hanna and a bit further—then it turned around and flew back up the river, right by our creek again, lower than the end of the ridge I'm on. It continued on up the Charley, out of sight."

Garrett had made up his mind that he would stay on that ridge until someone came to get him. "I have food for well over a week, maybe two weeks, and everything else I need. Walking to the mouth of Charley River or Coal Creek is possible but maybe a 3-day hard walk. Coal Creek mouth is about 25 miles straight line and about half high ground above timber which is easy walking. But I don't know what I'd do if I got to the Yukon at Coal Creek. Probably nobody there. Charley mouth is about 35 miles straight line and much more forest to go through."

Garrett knew well what thirty-five miles in spruce muskeg could be like. Sphagnum moss in this country can grow ten, fifteen-inches deep and blanket all manner of hazards underfoot. It is the kudzu of the north, and with the decomposer organisms here lagging by decades in breaking down old stumps, deadfalls, and roots, sphagnum has plenty of time to smooth over the pitfalls, creating a soft, spongy, benign illusion. And what sphagnum moss doesn't hide, a veil of wispy horsetail, growing thick and knee deep in places, does. Bushwhacking in the spruce lowlands in summer is simply miserable. Unable to see the obstacles, you trip, lurch and stumble your way around, banging shins and knees, twisting ankles and pulling muscles until exhaustion. Garrett also knew that a thirty-five mile straight line would be closer to forty-five miles by the time he dropped down into and then climbed out of the steep drainages that dissected this land. "No," thought Garrett, "a helicopter must come here within a few days. The best thing for me to do is to wait here patiently, eat well, stay close to camp, and stay healthy. And find some way to occupy the time!"

On the fourth night it rained hard.

"Rain started about 4 pm, light at first. . . intermittent. . . then steady at 6:30 pm. Colder now—total sky is dark and filled with rain clouds. I hope and pray the helicopter gets here tomorrow. I fear for the other 3 men and pray they are alive through this night."

"Rain stopped at 8 pm. Dark clouds all over. I lit the lantern and set it out on the rock once more."

"Rain again—harder."

Maybe we were cheechakos—the not-so-endearing local term for greenhorns—by Alaska standards, but I was not without woods experience and neither was Eduardo. From childhood, I preferred to be outdoors. In the Berkshire Hills of Massachusetts, I learned the names and uses of all the trees and most of the wild plants while I was still in grade school. Before I was a teenager, I realized my earliest dreams of building a log cabin. It wasn't much, looking back: just an eight-foot-square room of cribbed white pine logs, stacked with the help of my brother Philip and a friend across the street. But it had a waterproof pole-roof covered with tar-paper, a door, a window, a table, and it was in the woods. After school we'd steal off to our secret location and there we made bows, arrows, and spears, and hunted anything that scurried up trees, flew, or croaked and jumped into water. One day I almost hit a duck rising from a water hole, my arrow passing right under its feet. But that was about as exciting as it got. We rarely caught anything. Then I was given my first air rifle and my play in the woods turned deadly for chipmunks. I became a crack shot.

Winter was my favorite season, and by the time I was a teenager I ran my own trap-line in the woods nearby—a short one, for muskrat mostly, but now and again an ermine or opossum, and one day a skunk. I'd be out of the house early on winter mornings, full of anticipation, checking my traps in little over an hour so that I was back in time for breakfast and the school bus—except for the day I caught a skunk. Trapping for me, though, was only an excuse to be in the woods. I soon found I didn't have much stomach for it. An animal that wanted its freedom so badly as to chew nearly through its own leg to escape made an impression on me. Using leg-hold traps meant I had to kill the captured animal by hitting it over the head, or holding it under water until I saw no more air bubbles. Bashing the animal didn't appeal to me and drowning took too long, so I bought some new quick-kill Conibear traps and tried them, but the traps needed to be submerged and the water-courses I was working were too shallow. Still, I learned more about animal ways during those two or three winters than I had learned by any previous experience.

When I was old enough to have a hunting license I bought a used Mossberg bolt-action shotgun, sixteen-gauge with adjustable choke, and proved a poor marksman with it. My father, who didn't hunt himself, would walk with me through the old apple orchard behind the house and when we flushed the occasional ruffed grouse, I'd throw the gun up and shoot behind the bird every time. Then an old farmer in Richmond gave me an ageing rabbit dog, a small beagle with a hoarse voice named, appropriately, "Hunter," and I tried my luck on cottontails with no better success. One day a friend of my father's, with an eye on my dog, asked if I'd like to go snowshoe hare hunting up in Windsor— spruce-fir country. Almost as soon as we got into the woods, Hunter flushed a hare in sight of Mr. Ben. The two of us were fairly close together, but couldn't see each other in the dense cover, so Ben hollered before shooting to make sure I wasn't in the way. Only problem was he didn't give me time to shout back. As chance would have it the hare was coming straight in my direction, and just as I saw it, Ben fired. I caught the stray No. 8 shot with my mouth wide open. One pellet bore right through my tongue and into the back of my throat, but everything else missed my face and lodged in my heavy hunting jacket instead.

For my part the shotgun pellet did little harm. I walked into the emergency room, technicians took an x-ray and the doctor decided to leave it just where it was. For Ben, that was the last he hunted.

My next hunting adventure was with another of my father's friends. I didn't have a driver's license, so this fellow invited me to go deer hunting with him and in preparation we went to the local shooting range to sight in our shotguns with deer slugs. It was then I discovered that my old Mossberg had a misaligned barrel. At fifty yards my slugs were hitting the target in a nice tight circle—three feet to the right and three feet low. So I fashioned a rear sight-post out of coat-hanger wire to compensate for the defect, screwed it to the gunstock, tested it, and next day killed my first buck with one shot to the heart. That was the last I hunted.

Now I longed for the rifle. If we as a species aren't anything else, we are tool makers and tool users, and without our tools now we were a pathetic lot. We had done well enough in fashioning a shelter with a four-inch knife blade, the one advantage we had, but when it came to procuring food we were woefully unsuccessful. We tried building a rock-dam across the tributary creek flowing into the Charley with the idea that we could trap grayling in the small pool, then wade in and catch them with our hands, but there were no grayling in the creek. We scoured the area for signs of bark beetles, led by our own bark beetle expert, anticipating that we could pry the larvae out of stricken trees with our knife and make a meal of them. Bark beetle grubs are barely half-an-inch long, not much bigger than a pea when they curl up, but what they lack in size they make up in number—hundreds when you find an infested tree. But there were no bark beetles either. When the dog treed a red squirrel in a tall spruce that leaned out over a stretch of dry river bed, Eduardo threw rocks at it until he knocked it out of the tree, killing it. Three times in the next two days the dog ran a squirrel up that same leaner and each time we stoned it, but red squirrels are nothing more than fur and bones, and the amount of energy expended in stoning them far exceeded anything we got in return. Our third night on the Charley, we watched helplessly as a cow moose forded the river not sixty yards upstream. I cursed under my breath—as if not to scare off the moose—that all our firearms were with Garrett. And every morning when the sun first hit the high cliffs above us, we would see one or two Dall's sheep looking down on us; pure white specks balancing securely on the high pinnacles. They would appear only briefly and then vanish, hauntingly.

High bluffs on the Charley River, rendering smoke invisible and gunshots inaudible.

If there were something to be thankful for, it was that blueberries were starting to mature in the lowlands and required only hands to gather. On the bench of the Charley, near our lean-to, was a small bog with a fair number of bushes that were bearing fruit. More green than blue, and hardly filling, they at least occupied our attention as we foraged from time to time for handfuls of the almost ripe ones. Funny, I thought; when a person in the comfort of his home makes a considered choice to stop imbibing all but fruit and water for a period of time, and when that period has a planned beginning and end, it is called a fast. When things go wrong in the bush and a person is separated from his food and all customary means of obtaining it, and has nothing but river water and green blueberries to eat for an indeterminate period of time, it is not a fast.

The downside of blueberries was that the ripening fruits were likely to bring bears sooner or later, and in our case it was sooner. On the sixth morning, shortly after the sun hit us and we had all fallen fast asleep, a black bear came ambling up our side of the river. Bears are not known for having particularly good eyesight and the wind must have been carrying our scent the other way, for the bear was just about on top of us before the dog suddenly erupted in a terrifying tirade. In a complete stupor I found myself instantly on my knees, only a shadow's length from the bewildered bear, pointing and trying to shout something to Eduardo and Mark, but only mumbling incoherently. The bear seemed equally stunned and back-peddled a few feet with the dog alternately lunging at it and retreating in total fear. Then the bear stopped and stood its ground for a few seconds, assessing the situation with something like a 'what-the-hell?' look on its face. Accustomed to having its way out here, this bear could have done some serious damage right then, had it wanted to. But it either wasn't in the mood for a hassle or the dog was too much of an annoyance for it. It just turned its back on us and retraced its path a hundred feet or so, stopped and looked at us again, and then skirted around us through the middle of the blueberry patch in seeming indifference to our presence.

A couple hours later, I picked up an old tin can we'd found and wandered over to where the bear had detoured. I tried tracking it through the undergrowth just for something to do, following a faint trail of bent-over herbs, of broken twigs, of faint indentations in the soft moss. I picked a few berries whenever I encountered some that were less green. I poked around, thought a lot about where I was, soaked up the sunlight on my back. I came back to our campsite with half a can of blue-green berries and plopped down next to Eduardo.

"I don't get it, Ed."

He looked at me blankly. "Get what?"

"Myself. When I'm back home and stressed out, I dream of places like this, dream of just chilling out in the woods with nothing to do. So here I am in Paradise, even gettin' paid for it, and just 'cause I haven't got a cold Oly and a can of beans warming over the fire, I can't enjoy myself. I mean, look at this place. It's beautiful. If I had a gun and a pair of binoculars, I'd be lovin' it. Well, okay, maybe some of my sampling equipment too. And a sleeping bag." I was on a rant.

Eduardo looked totally bored. "Yeah, I've been having some of those same thoughts. I guess it's just not havin' anything to do. Dang it though, if I can't do my job, I'd rather be back in Seattle."

I rambled on. "Guess that's the difference between me and Fred. I have to be producing something all the time. I don't know where that comes from, feels like a curse sometimes, but that's how I got here, I mean on the Yukon, so it's not all bad. I couldn't do it the way Fred did. Too much a family man, I guess."

"Could you homestead here if you had your wife and kids with you?"

"I'd sure keep plenty busy. Fred is more hand-to-mouth, but I'd be growing and canning, curing hides and making clothes, working on a cabin, probably building my own canoe like I always wanted to. I wouldn't have to worry about producing, that's for sure. But then I'd probably be dreaming of moving to town. Would want it for the kids. And I do like big libraries."

I shut up for a minute, thinking about what I'd just said. Eduardo was quiet, his mind wandering back to Seattle. Finally I expressed what was probably on all our minds.

"What's worse than not having anything to do is not knowing how long we'll have nothing to do. Let's get ourselves outta here."

Day five had passed without our hearing a single airplane and it was quite clear now that no one knew where we were and no one was looking for us. Just waiting it out was getting on our nerves, to say nothing of our stomachs. It was time to make a decision.

Our topographic map was wearing thin, just from looking at it, it seemed, but all the staring in the world didn't change our situation. As near as we could tell, we were at least two drainages above Hanna Creek, maybe three, and probably fifty-five miles, as the Charley meanders, from the Yukon. To try to walk out would necessarily mean following the riverbank, as the canyon bluffs were too steep and the side creeks too deeply incised to consider short-cutting any of the river bends. It would be difficult at best; Eduardo and I already got a taste of river hiking over boulders, loose rock, and driftwood when we first hit the Charley six days earlier. And down on the flats the walking would be even tougher. The lower third of the Charley wanders like ribbon candy through extensive peatlands, which means that for miles we would either be slogging along river bottom or slogging through sphagnum bogs. We figured it would take another six days, and all that would do is put us on the banks of the Yukon. Depending on Mark's condition, it could take a lot longer.

Our other option was to build a raft and try to float out. Uprooted trees, courtesy of spring floodwaters, littered the riverbanks the length of the Charley. "Just as the salt water sends up its myriads of fish to feed the people of the interior, so the interior sends down its myriads of logs to warm the people of the coast"[1] With fire, we could burn trees into log lengths. Holding the logs together would be another matter, but belts, boot laces, and spruce roots might do it. Rafting would be tricky—uprooted trees also create log jams that we would have to muscle the raft around, and some of the rapids would likely be too shallow, or passages between boulders too narrow for a log raft. The river had come up considerably with the storms of the past week, but that was both good and bad. Higher water would get us over some obstacles, and sweep us into others. We had no rope by which we could line the raft through difficult stretches. We would just have to ride it out, using poles to maneuver through the tough spots. But navigating uncertainty was more appealing at this point than waiting for something to happen. If we could make it through the canyon, maybe thirty miles, the flats would be easy. Once on the Yukon we would still be eighty miles or so upriver from Circle, but that would only be a twelve- or fourteen-hour float through familiar territory. With luck we'd be back in Circle in a week.

We sat for a few minutes without talking, staring off at the river, letting the idea sink in. I turned and looked at Mark.

"What do you think?"

He shrugged. "Better than sitting here." The prospect of another week on the river was not something I expected him to rise to, but then it wasn't my preference either.

I looked at Eduardo.

Once we decided to take matters into our own hands—once we decided it was up to us to get out of this predicament—my morale lifted greatly.

The same was true for Eduardo. The same might have been true for Mark as well, had he not given in so completely to discouragement and resignation. We sized up some of the trees along the bank and decided to start with eight-foot logs of about ten inches in diameter, to see how it went. We singled out a couple for our first attempt and soon had two fires going under each. This would take a while, but we were *doing* something. Now all focus was on the task at hand. There was no other world, no outside, no family waiting; just the river, the logs, the fires. But apprehension was stalking my uplifted spirit. I knew this river could take our lives and I became haunted by two realities.

Mark was psychologically defeated. Though he was helping with fires now, he was devoid of any outward emotion. Maybe he sensed better than I the real danger ahead, but his lack of enthusiasm left me wondering how much he could be counted on if we got into trouble on the river. And as much as I fought it, this was becoming a source of frustration for me. While I had great sympathy for Mark, I hired him in part because I thought his physical strength would be an asset to our expedition. Such reasoning, however, was a measure of my own inexperience. What I failed to understand was that physical strength didn't count for much out here. Unless you are bent on wrestling bears, there is little about hunting, fishing, trapping, cabin-building, gold mining, or raft construction that requires an exceptional amount of brawn. The heavier tasks are accomplished with ingenuity. Out here, a clever mind and a sharp eye go farther than any three men trying to bull their way through life with brute force and dull tools. The strength to ignore mosquitoes, the confidence to be alone, the patience to observe and understand the ways of wildlife, the courage to face adversity—these are the qualities that make it here. And until this moment it didn't make much difference whether or not Mark picked blueberries, tended the fire, scouted for other edibles, or was present with us. But now faced with taking our situation back into our own hands, Mark's mental state concerned me greatly.

The other thing on my mind was even more troubling. As we moved about, gathering more fuelwood and muscling deadfalls around to position them better for our fires, we quickly discovered how little stamina we had. If we're going to have a fighting chance at this, I thought, we need to have something in our stomachs. I was looking at the dog.

I picked a quiet moment to broach the subject. We were sitting, tending fires, harboring our own thoughts and fears as the thin smoke spread out over the river. I got up from my fire, walked over to Eduardo, and sat down.

Eduardo somehow knew what was coming; it seems he'd been pushing the thought back for a couple of days, himself. I waited a minute, then asked softly.

"You thinking about the dog?"

"Yeah." He was staring at the ground now, scratching in the dirt at his feet with a stick.

I waited another minute.

"Think we need to do it?"

"Jesus it hurts." He snapped the stick in his hands. "He's done us well. He's gone through a lot these last few days, eating worse than us. Still he finds enough energy to tree a squirrel."

"I know"

Long silence.

Then I answered my own question. "Shit, I could never go through with it. What am I supposed to do, look him in the eye with a knife behind my back and say sorry ol' boy, you've been great, you kept a bear from walking over me, but we're hungry now? And I sure as hell don't expect you to do it. What's another fucking week without food anyway. I don't need it." I threw a rock hard at the log we were burning and get up. I was feeling anger now and I knew it was subconsciously directed toward Mark. He was the one I was damn worried about.

Garrett arose, fixed breakfast, then picked up his journal. "I lay in my sleeping bag until 7:15. It is a windless, clear day—bright sun warming things up already. Breakfast of bacon, eggs and hot chocolate over by 8:10. Then wash the dishes and wait."

The smoke from our multiple fires had settled low along the riverbed, creating an unusual aura of serenity this morning. Our new sense of purpose seemed to add to a feeling of calm. We stirred earlier than usual and with little conversation were up and about, gathering more wood along the riverbank and rekindling the fires of our hope. The tree trunks that we left smoldering all night had burned nearly through, and within an hour we had the first two logs for our raft. We chose three more downed trees close to our camp and started fires beneath them.

Our stamina was nil, but work was going well. Eduardo and Mark took a break to pick blueberries and I started scratching around in the shallow soil atop the riverbank for roots. Pulling up the organic duff and following fresh lateral roots carefully so as not to accidentally break them, I found that with patience I could chase roots of a quarter-inch or so in diameter out to six feet or more and end up with a good length of reasonably strong, pliable cord. It looked promising.

Mark was still moving slowly, still overly concerned with his physical state, but it seemed that his spirits were up a bit. He talked little, but was more present than he had been the last several days. When he and Eduardo returned with a hat full of blueberries, we sat down together against one of our logs.

The sun was well over the ridge now and its warmth permeated our bodies as we watched the smoke rise from our fires, scattered up and down the riverbank. Conversation was low-key.

"How many logs do you think we need?"

"Boy, when we get back to Circle, I'm gonna order up the biggest steak Frank has, with a pile of potatoes and one cold Oly."

"What do you suppose Garrett is doing now?"

"I don't know, but I'll sure be glad to catch up with him again."

"I pity the poor bastard, all alone, I don't care how much food he's got. Sure hope he's okay."

The conversation was punctuated by periods of silence, but the silence was more comfortable now. For my part, not thinking any more about the dog was a great relief.

I got up and went over to the nearest fire, poked it a bit and came back. We sat, contemplating; the hatful of blueberries unappealing. Another five, ten minutes went by without talking.

As I stretched back against the log, staring blankly into the clear sky, a vague awareness of something began to emerge from my subconscious, some elusory change in the air, in the flow of the river; something, I didn't know what. I turned slowly and looked at Eduardo, searching him for some sign of recognition. I turned to Mark and studied him. I was concentrating hard, sifting through the sounds of running water, imagining.

No. Maybe not imagining.

"Listen!" I half whispered. I was feeling something, more than hearing it.

I sat up.

"Listen!" I repeated with urgency, still softly as if not to drive away the feeling.

Eduardo's expression grew serious. He looked unflinchingly into my eyes, straining to hear above the constant rippling of the river.

We stood up slowly as if to rise above the background noise. Mark watched us. We stepped light-footed onto the open river bed and stopped, listening hard.

Then I could feel it again. Throbbing.

"A helicopter," I whispered. I looked at Eduardo. "A fucking helicopter!"

"Where is it?"

The sound would drift in and out. Faint. Far away. Then nothing.

Then the throbbing would return, a little stronger, only to be carried away again by the air currents or drowned out by the river.

"Jesus," I cursed. "Can't somebody figure out we're here? Can't somebody fly over this goddamned canyon and see us? For Christ's sake, how much more smoke do we have to put up?"

We were tense now, straining with every sense, frustrated that we couldn't hear better.

The wind seemed to change and the sound came back.

"I think it's getting closer! I think someone's coming!"

Mark was standing now. The dog's ears were cocked at us.

"YES! It's gotta be coming down the Charley!" We were shouting now, searching the sky upriver.

"It's coming! It's coming!"

"THERE!"

The helicopter exploded over the tree tops right in front of us, Gordon at the controls and Garrett next to him, the two looking down from that bubble in total disbelief. Mark clambered across the rocks and the three of us turned and grabbed each other, burying our faces in each others' shoulders, shouting, dancing, the dog barking amid the sudden noise and confusion. Gordon eased the helicopter down and Garrett jumped out, rushing headlong into the pack. It was mayhem. Now Gordon was out of the helicopter and I broke out of the pack to tackle him, smothering him with a hug. Ed followed, then Mark.

Gordon was stunned. He pushed himself free and ran back to the helicopter, wrestled a metal box out from behind his seat, and scrambled back across the rocks to us. The box contained survival gear, including a handful of chocolate bars, and Gordon shoved one into each of our hands. We ripped them open and stuffed them into our mouths, mumbling without pause. Everybody was jabbering at once and nobody heard anything because words just didn't mean a damn right then. I grabbed Garrett again, bleary eyed, clenching his shirt with both hands.

"Jesus, Garrett. For such an ugly old coot you sure look beautiful to me now. Where the hell have you been, anyway?"

Gordon had no idea what kind of a situation he was flying into. In Circle, he and Frank had packed a couple of boxes of groceries for us and Gordon loaded them into the cockpit of his helicopter with satisfaction, knowing we'd be pleased with the fresh goods. Frank had buried a couple six-packs of cold Olympia beer in one of the boxes, with last week's newspaper stuffed around it hoping to keep in cold until that evening. As an afterthought, Gordon grabbed a couple of magazines from the trading post and threw them on top of the groceries. He sat for a few minutes on the empty cargo rack, sipping at a steaming cup of coffee and soaking up the morning sun. As soon as the mail plane was in, he thought, he'd check to see if there were any letters for us and be on his way. Mike told him a week earlier that he had moved us to the summit of Twin Mountain, noting casually that we had become separated but would find each other with no problem. "Oh, and by the way," he added just as casually. "They want to be re-supplied sometime around the middle of next week." Gordon took the "middle of the week" to mean Wednesday, not giving any more thought to our original request for re-supply in five days. Everything was going according to plan.

When Gordon landed at Garrett's campsite, he wasn't at all prepared for what he found. Garrett, believing the three of us had perished, had worked himself into a dreadful state of mind and had already started composing a letter to our families. As Gordon stepped out of his helicopter he was met by a panic-stricken, nearly unintelligible woodsman who shook him and babbled effusively with details of an airlift gone very bad seven days earlier. Gordon was taken completely by surprise. The two of them stashed the boxes of supplies in Garrett's tent and climbed into the helicopter, Gordon cursing all the way. "I can't believe this. That son-of-a-bitch Mike just sent me a bill for $1,400 dollars. The FAA will have his ass but good for this and I'll make damn sure they do." He was mad and he was scared. His immediate objective was to contact the Alaska State Police to initiate a search.

Still basking in the moment, we calmed down enough to do some quick strategic planning. Gordon could carry two people with him back to Circle and there was no question that Mark should be one of them. I didn't want to subject him to the stress of waiting for a second pick-up, of having to watch Gordon fly out of here without him on board. Nor did I want to send Mark and Eduardo together. They had seen enough of each other for a while. So Garrett and Mark took the first shift while Eduardo and I stayed on the river. Gordon left us the remaining two chocolate bars and promised he'd be back in three hours. As he lifted out of there, Eduardo and I just turned to each other and smiled great big relieved, victorious grins. "Thanks, buddy," I said, and gave him another hug, slapping him on the back. I stood on the riverbed, scratching the dog's ears and looking around at the beautiful canyon we had found ourselves in.

In little more than two-and-a-half hours, Gordon was back. He emerged from the cockpit carrying three Cokes, a bag of potato chips, a handful of dog biscuits that he had grabbed hastily from the trading post. The three of us just sat in the sun with the dog and talked, no longer in a hurry to go anywhere. Eduardo and I had broken up our fires, spreading them and dampening them down as best we could using Eduardo's hat and the old can that we found to carry water from the river. An afternoon thunderstorm would take care of the rest; either that or start something a whole lot bigger, in which case our fires would be inconsequential. But the ground was moist from frequent rains the past week and we felt comfortable leaving the ashes now. We were ready to go.

On our way out, flying just above treetop, we encountered a large grizzly, the largest Eduardo and I had yet seen, not two miles below our

campsite. It was working its way slowly up river. Alarmed by the helicopter, the bear broke into a full gallop through a stretch of open spruce
muskeg in a chilling display of speed and power over spongy, sphag-
num-covered ground. As we overtook it, the grizzly pulled up short to
make a stand. It reared up on its hind legs, frantically clawing up at
us, mouth salivating and long yellow-brown fur rippling with fearful
strength. I'm not usually given to negative thoughts when I'm in an aircraft, but for a split second I had a vision of us stalling. I shuddered as
Gordon maintained his flight path. Glancing back at the bear, the three
of us shot looks at each other that needed no words.

I didn't eat as much as I thought I would back in Circle. One of Frank's
bacon-cheeseburgers, followed by a shower and shave, satisfied me
thoroughly. After a comfortable night's sleep in a bed, I felt quite good.

In the morning, two tasks occupied my thoughts right away. One
was to make arrangements to get Mark started on his way home. As
it turned out, Frank and Mary's daughter would be driving into Fairbanks later that day and agreed to give Mark a ride. Mark was glad not
to have to get in a small plane again and I was relieved that he could
be on his way out so quickly. In Fairbanks he would do what he could
to rebook his return trip and, hopefully, be headed for the East Coast
within a day or two.

The second task was to retrieve our gear from Twin Mountain. When
Gordon found Garrett out there alone and heard his story, they left
everything right where it was in order to get back to Circle as quickly
as possible. It was only because Gordon chose to fly down the Charley that they found us. When I asked Eduardo how he felt about going
right back out, he suggested tactfully that he was more than content to
stay in Circle for a couple of days and catch up with his writing. So Garrett and I asked Gordon to fly us to the Lonesome Ridge campsite—the
name bestowed upon it by Garrett—that afternoon. Since neither Garrett nor I had much interest in Circle, we decided we would stay on the
mountain for a couple of days and scout the area. We'd take a minimal-
ist approach and send out as much gear as we could with Gordon on his
return trip. He'd be back in two days to get us and the rest of the gear.

VI.
HEADWATERS

THIEVING PIKAS AND SINGING VOLES + FORTYMILE
CARIBOU + TALES THE TREE RINGS TELL + TRUTH IS
A CHAMELEON

From the air, the big tan-colored, four-person tent could be seen a mile out, standing alone and alien on the north ridge of Twin Mountain. Everything appeared undisturbed since Garrett's hasty departure twenty or so hours earlier, allaying my greatest concern. Food and gear can be replaced, but if a bear, tearing up a camp, finds even the smallest reward, it could mean trouble for anyone in its territory.

Gordon circled the campsite and set the helicopter gently on the tundra, slightly uphill from the tent. The tent rippled and leaned in the blast of the rotor, but stood well. Gordon killed the engine and as the rotor wound down we sat silently for a long minute, in the company of our private thoughts. Each of us had a different emotional connection to the place, each strong. For me it was the answer I had been seeking ever since we became separated on that stormy night, something I felt at times desperate to know. Where was Garrett? As I took in the enormity of the scene, my first thought was of the impossibility of our situation: how absolutely alone Garrett had been up here, and how totally improbable it was that we should ever find each other.

I took a deep breath.

We climbed out of the helicopter and walked over to the tent. Gordon sized things up. "Yeah, we're good here. If most of this stuff can go

Garrett's "Lonesome Ridge" campsite.

back now, I'll have no trouble with just you guys, the tent and whatever else you have when I come back for you."

I looked at Garrett. "Maybe we ought to keep four or five days worth of food in case the weather closes in, eh buddy?"

Garrett just bobbed his head a couple of times, with a hint of incredulity in his expression. "Yeah," he said, as though he couldn't quite believe we were doing this.

I picked up the rifle.

"Keep this?"

"No, I don't have any shells left. Let's keep the 12-gauge and the other two can go."

"Pots?"

"Maybe just one for water and one fryin' pan. We can do simple stuff for a couple of days."

"Okay, Gordon. Those other tents, Mark and Eduardo's sleeping bags, pads, extra stove, Ed's insect traps, guns—we don't need the lantern, do we Garrett?—all that can go back with you now. Garrett, you wanna pick out some food? Let's keep all the eggs, onions, potatoes. And cheese. Man, I'm craving an omelet."

An hour later, Garrett and I watched Gordon fade toward Circle. I turned and looked at the sky behind us.

"Clouds building already, Garrett. Wha'd'ya think? Guess we'd better get to work, eh?"

"How about some lunch first?"

"Won't turn it down."

I felt good, glad to be in the high country. I looked around at the widely scattered clumps of spruce forming tree-islands around our camp. For the most part they were growing upright, unlike the gnarled, matted, wind-blown krummholz commonly found at treeline on so many mountains in the West. Many exhibited dense foliage at their bases where winter snow cover afforded some degree of protection, while just above the average snow depth, branches were trimmed by the abrasive action of wind-blown ice particles. Once the tree grew out of this zone of abrasion, it again assumed a straight and symmetrical form with robust foliage. The shapes of these trees already told me a lot about growing conditions here, but coring them would tell even more and I was eager to get started.

Garrett, too, was scanning the area as if seeing it for the first time. Alone and desperate here for seven days, he recorded not a single observation in his journal beyond weather conditions, his repeated efforts to find us, and his fears. That told me all I needed to know about his mental state, as Garrett was normally prolific in his note-taking. Standing next to him, I slapped him on the shoulder.

"Let's eat."

We picked a couple of flattish rocks to sit on, cut up some cheese, opened a can of sardines. Life was good again.

We talked little, still reflecting on the events of the past week, glad to be together. Garrett finished the last sardine, stood up, and wiped his hands on his pants. He walked over to a pile of gear lying outside the tent, picked up a sack full of traps and brought them over, and sat again. He dumped the traps on the ground between his feet and counted them into two piles, sixty for each of two lines. He stuffed the first pile back into his canvas bag and walked up beyond our landing site to a rocky outcrop. After inspecting the ground carefully, he placed several traps under shrubs and in rock crevices. With the tenth trap set, he straightened up, eyed the ridge for a minute, and then paced off a hundred feet to a second station. The process repeated itself. Another hundred feet, inspect the ground, ten more traps, and on he went until he had all sixty out. At each station he placed traps wherever he found scat or small runways in the tundra vegetation, and it seemed to him there was no shortage of either. In two hours time, he was starting a second line from the tent down, in moister tundra covered with sedges, sphagnum moss, dwarf willow and alder thickets, and now occasional spruce clumps.

While Garrett was putting his traps out, I hiked off in a direction parallel to the ridge to a group of larger tree-islands to set up my own sample plots, where I would count and age seedlings, and core the larger trees for more complete tree-ring analyses back in the laboratory. My eventual goal was to establish a series of plots running downslope some distance from treeline in order to determine how forest age structure changed with elevation near the upper limits of growth. Most of this work would have to wait until we set up our next camp, though, since Eduardo would be with us then and we would stay longer. For now I was content to establish a few plots, test my sampling plan, get a feel for what was going on, and see what modifications might be required.

I became immediately engrossed in my task, laying out plots ten meters on a side, crawling on my hands and knees through the miniature heaths of the tundra mat, looking for tiny tree seedlings; then fighting my way into the dense centers of spruce clumps to obtain increment cores from the larger trees. So absorbed was I that I had completely lost track of the ominous cloud reaching for the stratosphere above me. When the first lightning bolt out of that towering anvil hit the ridge we were on, I jumped so hard that I ripped my shirt on the branch stub next to me. I pushed backed out of that spruce clump wheeling and stumbling, glanced up and then looked over to where Garrett had been working. He was already scrambling up to the tent, hollering something my way and waving. I dropped everything, not caring to make a run for it with a steel increment borer in my hand, and started to sprint.

"Whoa, wait a minute." I turned back to grab my field notebook and the few cores I had already collected, and bolted again for the tent. The rain caught me two-thirds of the way there and Garrett was holding the screen open for me as I half dove, half rolled in.

"Jesus, that came up awful fast didn't it?"

"Yeah, it snuck up on me, too. I had my nose to the ground. Man, I was finding all kinds of animal signs out there"

"Gotta get out of this shirt, I'm c—"

BAAM!

I couldn't get the sentence out before all hell seemed to break loose around us. Suddenly it was one blinding lightning bolt after another, hardly a pause between them. With each one we ducked reflexively, as if by so doing we could dodge a direct hit. The salvo was overwhelming and we were helpless. We had no escape, no place to run. Now I was scared, almost panicky. Never had I experienced a moment, not even when our pilot screamed at us eight days ago to jump out of his helicopter, that I could remember feeling like this.

"Jesus, Garrett," I shouted above the thunder. "I don't like this one fucking bit! Look at us, for Christ's sake. On top of a mountain, we're five times taller than anything around us, and these tent poles are nothing but a goddamn lightning rod!"

BAAM!

We flinched in unison.

"Enough! I've had enough!" I cried out to no one, and threw myself down on my sleeping bag. I buried my head and pounded the ground rhythmically, pushing back my fear. I was no longer in control—not of what was going on around me, not of myself. My nerves were shot and I just lay there now in complete submission. Garrett remained silent in the crashing thunder. Then, just as spontaneously as I had broken, I calmed. My mind emptied of all thought and I became completely relaxed, like a person with hypothermia, just before dying. I felt extremely tired. The flashes and the thunder became surreal, dreamlike.

I must have fallen asleep. When I raised my head Garrett was lying on his bag next to me, reading.

"What time is it?" I groaned.

"Six-thirty."

"Has it stopped raining?"

"For the most part."

I listened. The sky was still rumbling, but the storm had moved to the east and the sun was trying to break through the turbulent clouds in the west.

"I'm hungry."

"Yeah, me too." Garrett rolled onto his side and put his book down. "I was waiting for you to wake up."

"Jeez, that was something."

"Yeah. You okay?"

"Yeah. Sorry I lost it. Guess we're still alive, eh?"

"Let's see what we can find to eat."

When the rain finally subsided, we stepped out of our tent to have a look around. The sky to the east was heavy and turbulent. Someone up toward Eagle was getting it pretty good, but the sky over us was breaking up now and the tundra was sparkling in the low sunlight. The air was delicious with the fragrance of alpine Labrador tea, a dwarfed relative of the lowland shrub bearing the same name. Like creosote bush in the desert Southwest, Labrador tea releases a volatile compound when wet, giving the northern landscape a characteristic redolence after rain. So pleasant is it that I have often kept a handful of dry Labrador tea leaves in a small birch-bark basket in my study, and every once in a while I would mist them with water to invoke nostalgic feelings for the Far North. As I took in the scene now with all my senses, I felt very much alive; almost immortal, as if the turmoil of the last few hours had completely recomposed me.

We climbed up on the rock where Garrett had kept his lantern flame burning for us through the last week, and there we presided over an immense landscape. I found myself repeatedly turning toward the east, toward where Mark, Eduardo, and I hunkered through our own stormy days, tending our fires and hopes beneath our birch-bark lean-to. I tried to visualize our smoke rising lazily from a distant canyon, tried to visualize a silver-winged airplane flying low over the river to inspect the men down there. I could see the men sitting around a fire, looking to the sky, talking about building a raft, but they were faceless. They didn't seem to be us.

Garrett spoke first.

"I covered a lot of that ground last week" he said thoughtfully, having just gone over all the terrain again in his mind.

"I guess you did." Pause. "There sure is a lot of room out there."

"Can't figure out what Mike was thinking."

"We tried for days to second-guess him. Finally gave it up."

"Guess Gordon will get some answers. He got pretty scared when he showed up here and I told him what happened. Now he's fightin' mad over this whole thing."

"Good. Let him go after Mike. I'm tired of thinking about it. I'm just glad to be here right now."

Then I changed the subject.

"You gonna check your traps tonight?"

"No. That storm would have kept the critters down. I'll stay away for now and check 'em in the morning."

After another minute I spoke again. "Things are gonna change, Garrett. I can feel it. I'm really looking forward to the next couple of weeks."

"Yeah. It'll be good. I'm liking it a lot better up here now that you're back."

After a quick meal of pilot bread, Nabob jam, cheese, and hot chocolate, we set up our spotting scope to have a look around in the late sunlight, hoping to catch sight of a caribou or grizzly in the distant tundra. This was our recreation. Far to the southeast, on a steep slope falling away to the Charley, a pinpoint of light shined brilliantly. Too small to make out even with our 40X scope, it appeared to be a reflection off something metal. Remnants, we guessed, of the B-24 that crashed at the headwaters of the Charley in 1943, four days before Christmas. The story of that crash and of the lone survivor who managed to walk some 120 miles out the Charley River in the dead of winter, is part of the local lore; not so much for the woods skills of the airman—he apparently didn't have any—but for the indirect help of a local character, himself something of a legend, and the unwritten bush code that cabins are to be left unlocked and the elevated cache stocked with basic necessities for anyone whose life might depend on them.

The story has undoubtedly been embellished over the years, but the essence of it is this:[1] First Lieutenant Leon Crane, co-pilot of the stricken bomber, had parachuted safely from the spinning aircraft and was the only one of the five-man crew to make it. (As he was descending he saw one other chute above him, but the chute disappeared on the other side of a ridge and the airman was never found.) Crane had a rough time of it in the deep freeze for his first week. He was outfitted for winter flying, dressed in a hooded, down-filled flight suit over long johns, was wearing felt-lined boots, and carried matches; but even so, forty or fifty degrees below zero in the dark of winter remains dangerously cold. He wrapped his parachute around him, built a fire, and kept himself going for a while, gathering firewood with cracked and bloodied hands while dreaming of hot food. He stayed put for nine days on that steep slope where he had landed, but finally, realizing that no one was coming for him, decided to start walking downhill even though he had no idea where he was.

Crane plowed his way through the deep snow for a full day before he stumbled upon a cabin on the Charley. The cabin belonged to one Phil Berail, a tough trapper and miner from Woodchopper,[2] and was well stocked. Berail had put up plenty of dry firewood and the cache contained food, clothing, a rifle, ammunition, and snowshoes. So Crane settled in, ate generously, slept warm, and nurtured himself back to full strength. Comfortable now, he was in no hurry to leave the cabin. He stayed there forty-three days before deciding to make a go of walking out. He had planned to wait until spring, build a raft, and float out, but apparently thought his, or rather Berail's, food would not last that long. So he fashioned a sled with boards and a wash tub, stocked it with food enough for two weeks and headed down the Charley. After fifteen days of laboring in unpacked snow, he came across another cabin with food—maybe the one we missed—and stayed there three days before continuing. A week later, seventy-eight days after parachuting from an out-of-control bomber, the airman showed up at Al Ames' cabin on the Yukon. Ames took him downriver by dogsled to Woodchopper, where Crane met and thanked the sixty-five-year-old Berail. The next day he was flown out by bush plane.

The story is remarkable in many respects, but it leaves something unsaid. Crane was young—twenty four—and unmarried. As an enlisted airman, he wasn't going anywhere for a while, nor did he have to worry about reporting to his senior officers. For the time being, at least, he had no responsibility to anyone: he was presumed dead. He had suffered greatly for nine days, but then found shelter, even comfort on the

Garrett examining one pika's hay pile.

upper Charley, courtesy of Phil Berail and the ethics of local woodsmen. But he was alone, and all the while he was out there he had no
way of knowing the fate of his four crew members, of the parachute he
saw above him, of what the others might be doing, whether anyone was
looking for him or had any idea that he was still alive. This was his tor-
ment, as it was Garrett's. Neither talked about it.

The next morning dawned clear and noticeably cooler; not quite freezing, but getting closer. The signs were subtle yet, but fall was already
coming on in the high country. Garrett caught four red-backed voles on
his lower trapline and all were showing increased deposits of fat, especially in the region between the shoulders where brown adipose tissue
was accumulating. This was the fuel that would take them through
the coldest nights when food was scarce. Garrett thought their fur was

considerably thicker, as well, compared to animals he had caught ten days earlier on the Nation River.

The pika haypiles, too, were getting larger by the day. Garrett was fascinated by the habits of these industrious little animals of rocky alpine slopes, and between his trap stations he would stop and watch them. Much of the pika's time was now occupied by running back and forth between their home territories in the rocks and the tundra vegetation beyond, where they would clip mouthfuls of plant material for their winter stockpiles. They were selective about what they gathered, too. Plants harvested for winter stores included a disproportionate amount of material that the pikas would not consume in their summer diets: plant species high in chemical compounds designed for the very purpose of deterring herbivores. It was a curious choice, but researchers eventually discovered that the phenolic compounds serving as plant defense in these species also inhibited microbial decay of material in the pika's stockpile, and by mid-winter the distasteful compounds themselves would degrade to the point of palatability. So the pikas were, in effect, planning for the future, ensuring that a portion of their stockpile would be preserved until needed late in the winter.[3]

It was a busy time for the pikas. Whenever they ventured out into the tundra, they endangered themselves to predators, mostly weasels and hawks, so they would work fast and scurry back to the rocks with each mouthful, which they tucked out of the weather to dry under overhanging rocks. There was another reason too, to keep moving. The pikas also had to guard their stockpiles against thievery by less industrious individuals. Because their habitat preference is so specific, many pikas live in close quarters within a relatively small area of rock talus, where each maintains and defends a specific territory. But trespass is common, especially when the neighbor is off gathering plants, and thievery of another's food stores is not above pika morals. I have known biologists to spend countless hours observing the pikas' food-caching behavior and watching the interactions between neighboring individuals to determine the energetic payoff of putting hay up versus stealing it. The pika that steals from a neighbor spends fewer hours foraging, exposes itself less to predators, and enjoys the benefit of its neighbor's food-selection effort—advantages that are not lost on the pika. When researchers put out a pile of clippings in undefended territory it was often stolen within hours. But the thief of a neighbor's hay runs the risk of severe aggression, so successful robberies are kept to a relatively low level, usually involving only a few repeat offenders in the population.[4]

Half-way up his second trapline now, Garrett had picked up three more voles. I was just finishing camp chores when he suddenly broke the silence, yelling down to me and holding up something I couldn't see. I started quickly up to meet him as he was coming fast toward me.

"We got one!" He shouted. "We got a singing vole!"

The vole was dead as could be, but Garrett had his hand wrapped around it like it might still get away.

"Look here," he said when he got to me. "Isn't it beautiful?" He launched into a monologue.

"See how short the tail is? Look at it next to this red-backed vole." He reached into his pocket and pulled out a cloth sack with the other voles in it. "And see how reddish this fur is? Now look at this one." He pushed the newer specimen at me, holding it in the sunlight. "The singing vole is just this nice brown color on the back, and buffy here on its flanks." Garrett was handling the vole now like he'd just picked up an interesting rock, rolling it from one hand to the other, looking at it from every angle. "Oh man, and look as these claws. They're a lot longer than in any of the other voles I've been catching."

Garrett had reason to be happy. It was a good find. Because of its isolation on mountain tops in this extreme northwest corner of the continent, little was known about the singing vole except that it exhibited certain behaviors that were more like that of the pika, with which it seemed to share a preference for rocky alpine habitat. Like pikas, these semi-colonial animals are rather vocal, often sitting on rocks or other exposed places and emitting a metallic churring sound, for which they have been given their name.[5] And like pikas, singing voles store plant material in hay piles for winter, though there has been talk about this being done primarily by juveniles, possibly involving a cooperative effort among siblings. But much about this behavior in singing voles is speculative and Garrett would have plenty to think about now. He wondered aloud if there was something about this environment that selected for similar behaviors among the two unrelated species.

Revelation in science, whether a grandiose concept or a small detail, always carries a price. Bucking conventional wisdom means subjecting yourself to the greatest scrutiny and criticism, even cynicism, of your peers, some of whom will have a vested interest in the convention and may be quite unwilling to listen to alternative ideas. I had the rare opportunity to experience this even before I was awarded my degree. Having come from the physical sciences, where mathematical modeling often led the way to fresh thoughts about old problems, I applied my training as a newcomer to the biological sciences. Examining a long-revered concept relating to the growth of trees at their elevational limits, I surmised that the dogma of the times was wrong. I soon found myself very lonely.

Treeline is one of the most abrupt of ecological boundaries on our planet (shorelines being another) and the phenomenon has attracted the attention of scientists for more than a century. Among the various conventions that have floated to the top of the literature is the notion that tree growth at high elevation is limited by excessive moisture loss, especially during the winter months, resulting from exposure to high winds. When I saw a problem with the rational behind this notion and tested conventional wisdom in a wind tunnel as part of my dissertation work, I upset the apple cart. With the publication of my first paper on the subject, I drew polite but intense criticism, mostly from forest ecologists in Europe. The idealistic side of me knew this was good for science—this continual re-examination of our perceived truth is the process by which science works—but it can be rough on the individual until the challengers, one by one, are quieted. It's a good idea, I learned, to be sure of the ground on which you are standing before firing the first shot.

Now, quite unexpectedly, and in a place hardly at the forefront of modern science, I was facing prospects of upsetting another convention. The trees I started coring yesterday before the storm—the clonal islands that established the upper limits to tree growth here—were white spruce, not black spruce as in eastern Canada and the northern Appalachians where I hailed from. The prevailing literature held that, unlike black spruce, white spruce did not reproduce by cloning. Black spruce in the northern muskeg spreads vegetatively by developing adventitious roots from branches that touch the ground. That in turn alters hormone distribution within the branch, turning it upright so that it grows into a trunk with branches of its own. We know this because a number of investigators have observed it and reported it in the literature. So, we expect this of black spruce. But we know also that white spruce does not do this because nobody had observed and reported it in the literature. And if that were not troubling me enough, the first few cores that I extracted were all wrong as well. My initial examination of the annual rings showed a clear pattern of increasing growth over the past several decades, completely the opposite of what I expected and what I observed on the slopes above the Yukon. This wasn't supposed to be, either. I had that lonely feeling again, now questioning my own understanding of the literature, questioning my ability to differentiate between the two species of spruce, cursing my ignorance. There were no experts to consult with now. I was alone.

I worked quietly through the day, deep in thought.

"How's it going?"

I jumped as Garrett startled me out of my concentration. He had finished his work for a while and hiked over to see what I was up to.

"Jeez, I'm glad you weren't a bear!"

"Yeah, the ground is quiet. Nice and soft after the rain."

"It's going okay, I guess. I'm pretty puzzled over something here, but I'm just gonna go ahead with my sampling plan and see what turns up."

"Anything I can do to help?"

"Well, sure. I need to measure the basal diameters of a whole lot of these trees. It'd be a lot easier if when I fight my way into these miserable clumps with my tape I could just shout out the measurements and have you write 'em down."

"Glad to."

At camp that evening, Garrett prepared his study skins as he usually did, spending a little extra time over the singing vole, and I tidied up my field notes. A pot of hot water steamed on the camp stove and we refilled our mugs several times with tea, savoring the camp life again. As the faint sound of an aircraft began to register with me, I looked up from my notebook and gazed absentmindedly toward the west. Planes are few enough in this country that it's hard not to look for it when you hear one, especially when you were sitting on top of the world. The sky to the west was troubled again, clouds building and the wind picking up a little.

In a few minutes the sound clarified, and I spotted the small black dot against the cloud bank.

"Helicopter coming," I said quietly.

Garrett looked up from his work, oblivious to the sound until then. "Gordon?"

"Don't know who else it would be."

We both studied the black dot as it grew larger.

"Were we expecting him?"

"Not 'til tomorrow."

I took another sip of tea as I watched Gordon come straight in. He slowed to a hover and set his machine down gently about forty yards away. Neither of us got up until Gordon cut the engine and climbed out of his bubble.

I hid my curiosity with a grin. "Hey, Gordon, you bring us a cold beer?"

"Don't you wish. How're you guys doing?"

"Great. What's up? I'm pretty sure you're not out here on a joy ride."

"Weather's turning. Looks like we're gonna get socked in for a couple of days. I was afraid I might not make it out tomorrow, so figured I'd better come out and see you guys now, see if you want to hunker down here or come back in to Circle."

"How bad does it look?"

"They're talking about a real low ceiling—couple thousand feet, which was my primary concern—and possibly heavy rain. Could get pretty windy, too."

"Well damn. That doesn't sound like a lot of fun." I looked at Garrett. "What'd'ya think? I've had a good day and don't need a repeat of yesterday, but I'm game if you want to stay."

Garrett didn't need much time to think it over. "No. I've seen enough here. Seen enough of the inside of this tent, too. As long as we can get back into the high country after the weather blows by, I'm happy to pack it up for now. Think we can stay in one of the bunkrooms?"

"Don't know why not. Can't imagine there are a lot of tourists in Circle right now."

It was a good call. The next two days were all that the forecast had promised. Rain beat a steady staccato on the tin roof of Frank's bunkroom, the wind occasionally rattling the loose sash, water droplets bumping each other down the dirty window panes and squeezing through the old, cracked glazing onto the inside sill. We lay on the bunks, writing a little, napping a little, until the chill would get to one of us and we headed for the hot shower for the second time since being lifted off the Charley. Then the rain would let up some and we'd throw on a wool jacket and hat, and walk across the road to the Trading Post, buy a candy bar, walk down to the river. The rain would come harder again and we would beat it back to the bunkroom, glad not to be out cooking in the rain, not having to deal with personal business in the sopping wet bushes, not having to stay in our sleeping bags to keep warm. But at the same time we were restless. Eduardo had been in Circle two days already and had done all the washing and letter writing he needed to do, and I had gotten just enough taste of the high country to be itching for more.

Gazing mindlessly through the raindrops on the bunkhouse window, my blank stare eventually settled on Eduardo's truck, parked for the summer behind an old garage.

"Hey Ed," I said matter-of-factly, "It looks like your tires could use a little air."

Eduardo got up and wiped a circle of condensation from the inside of the window.

"Yeah," he said, "not surprising." Then he livened up a little. "Hey, you guys up for a ride? Why don't we get some air into those tires and take a spin up the Steese Highway."

An hour later, the three of us and the dog were crowded into the cab and headed up the road, windshield wipers slapping out their rapid beat while the defroster and I in the middle worked to keep a clear spot on the inside for Eduardo to peer out.

Eagle Summit is a gently rounded pass, slightly more than three thousand feet in elevation, but with a reputation befitting a mountain four times its height. The weather on top was considerably worse. Completely socked in, visibility was barely fifty yards when the clouds thinned a little, fifty feet when they closed back in, and the wind was driving hard. A carload of sane people would have turned around right there. But we needed to stretch our legs and we had the truck for shelter, with Frank's bunkhouse forty miles down the road to back us up, so we parked as far onto the shoulder as we could and went for it. Snugging our raingear tightly around our necks and pulling our hats down low, we slid out of the truck and started across the tundra, just to have a look.

When you are prepared for it, bad weather is never quite as disagreeable once you get out into it; or as a Swedish friend of mine was fond of saying, there's no such thing as bad weather, only bad clothing. Now that we were active, I found the wind and mist dripping from my face rather freshening. The cloud, too, seemed somehow to envelope me in a contemplative, mental comfort, and I was soon totally engaged in the mood of the place. The wetness of the tundra underfoot reflected the whiteness of the dense fog, imparting a light gray-green color to the ground that almost exactly matched the lichen-covered rock and graded seamlessly into the opaque atmosphere. Eduardo and Garrett walked on either side of me, separated by only fifty feet or so, and yet would repeatedly disappear in the drifting clouds and then reappear, vaguely, mysteriously, before vanishing again. Often we had little sense for the lay of the land around us. Sometimes the cloud would thin to reveal a sudden wall of rock; sometimes the cloud would thicken to mask an outcrop that wedged between us as we paralleled each other. On the leeward side of one such outcrop, a lone caribou appeared in the mist before me, almost as if one of my companions had been transformed, and neither the caribou nor I had any time to react before we were enshrouded again, and gone. So completely ethereal was the caribou, as I may have been to it, that I had good reason to later wonder if the experience was real.

It was a good walk. It was the first time since we came into the country that we had taken time to enjoy our surroundings without analyzing them. Our temporary grounding had its reward.

Circling around to the truck in two-and-a-half, maybe three hours' time, we drove back down into the shelter of the trees below, turned off the gravel highway onto a side road (the distinction between highway and road being a matter of about a car width, more or less), and followed it to an old mining camp at a place on the map called Miller's

House. As we were poking around there, curious about the trappings of the operation, a guy pulled up in a van who had lived there twenty-five years earlier when "Mrs. Miller served a great meal for practically no charge." Garrulous and happy to share his past, he recalled in great detail the workings of the camp and the characters who had hung their hats there for a while. He talked about hunting in the area, too, and told me that from time to time he had seen as many as twenty thousand caribou crossing the tundra at Eagle Summit during migration. It was a startling testimony to the changes that have occurred here since the early 1900s.

The caribou the man was talking about were part of the Fortymile herd, one of five major herds in the interior of Alaska, this one ranging from the confluence of the Yukon and Porcupine Rivers, southeastward along the Yukon drainage to the White River, and beyond. It was the same Fortymile herd that was nearly decimated during gold rush days, but which slowly rebounded after the wave of miners passed. By the 1920's the herd was back up to an estimated five hundred thousand animals, only to see its numbers fall precipitously again by 1940.[6]

Caribou herds are naturally prone to wide fluctuations in number, affected mostly by winter snow conditions and the changing fortunes of wolf packs in the vicinity of caribou calving grounds. But the Fortymile herd had had more than its share of trouble with humans as well. The only major roads in Alaska providing access to the Yukon River prior to construction of the pipeline haul road were the Steese Highway to Circle and the Taylor Highway to Eagle, both of which had been bulldozed square across the herd's migratory path near each end of their annual range. And both had become veritable firing lines for hunters that could now drive to their positions and wait in ambush for the herd. By the early 1970's the Fortymile caribou had shrunk to an estimated four thousand animals and no longer crossed the Steese Highway in their annual migration.[7] They had become so scarce that only five caribou were taken by residents in the area in the two years prior to our arrival.[8] My sighting of the lone caribou materializing out of the mist on Eagle Summit was only our second record for the summer.

While the diminished Fortymile caribou herd continued to move back and forth across the Taylor Highway at the east end of their range, the villagers at Eagle were also lamenting their low numbers. The native inhabitants were still dependent on caribou for their winter's supply of meat, harvesting ten to fifteen animals per household (until the early 1970's, almost no meat was purchased by residents from the village store in Eagle).[9] But it wasn't just the loss of meat that was felt by the locals. The winter coat of the caribou is one of the thickest of all northern mammals, keeping the caribou warm to minus forty and below without having to increase its metabolism to maintain body temperature.[10] The long guard hairs of the caribou are hollow and compartmentalized, providing exceptional insulation, and both natives and Anglos in the interior valued caribou for outer clothing and blankets; witness Fred's use of caribou hides for sleeping outdoors in winter.

We said goodbye to the stranger and returned to the bunkhouse in Circle. Twenty-four hours later, we were flying back into the high country, this time with our sights on the headwaters of Webber Creek, just west of Mt. Kathryn. Eduardo and I, with the dog, went out on the first lift to select a campsite. Garrett and most of our gear would follow on the second lift. We were particularly attentive to weight restrictions now and had scaled back considerably on what we took. Gordon could carry about seven hundred pounds plus himself, so his standard procedure was to top off his fuel tank, size up his passengers, and then load whatever cargo he still had allowance for. But I had already lightened my personal load significantly. Even before leaving our Kandik base camp, I had cut back considerably on what I found to be necessary and comfortable in the bush. Camp chair, hammock, reading light, battery lantern, axe; I'd already decided that I would leave this stuff in Circle on my way out. I had no use for these excesses now and would be glad to let others take them off my hands.

As we approached the upper reaches of Webber Creek, Eduardo pointed to a ridge that formed the divide between Webber and the Woodchopper Creek drainage to the east. Ceding to the engine noise in the helicopter, he looked back and simply gave me a questioning expression. I surveyed the terrain ahead and on the other side of the craft, and indicated my approval. Ed tapped Gordon and pointed it out, and with a nod Gordon veered toward the southern end of the ridge and a level area that was couched slightly below the height of land.

Alder Creek headwaters.

There we set down and quickly unlashed our gear without cutting the engine. Four hours later, Gordon was back with Garrett.

As we unloaded boxes of food, cooking utensils, stove, and all our collecting paraphernalia, Garrett verbalized his delight with our choice of site. And it was an extraordinary location. A five thousand foot ridge formed the southern horizon and marked the divide between the Tanana River and the upper Yukon (the Tanana spills off the north slopes of the Wrangells and runs a parallel course with the Yukon until the bigger river turns at a right angle at Fort Yukon and flows south-westward, absorbing the Tanana). Looking much like the glaciated Presidential Range in the White mountains of New Hampshire, the ridge sloped steeply toward us and was cut deeply by the headwaters of Alder Creek. The north face of the ridge was in partial shadow for much of the day, creating dramatic definition. To the west of us a lower ridge, smooth and emerald green for its sunlit tundra vegetation, gave us a window to the distant skyline of Alaska's own White Mountains. Indeed, this would be a fine work place for the next week.

We fell asleep to showers during that first night, but awoke to a beautiful clear morning and charged into our new routine. Right away, I spotted a distant site that I wanted to work: an isolated forest stand on a high, east-facing slope above the east fork of Alder Creek. It looked perfect for what I needed, a continuous forest stand extending over several hundred feet in elevation and reaching into the tundra with all appearances of a treeline on the move. Not one to make life easy for myself, the fact that the site would require a good couple of hours to reach was something I never gave a second thought.

"I see where I'm going today. Anybody want to check it out with me?"

Garrett opted to stay and start setting his traps out, but Eduardo was eager to get into the trees and hunt for bark beetles. He looked around at his options.

"Sure, I'm up for it." So the two of us packed up some food and rain gear, grabbed Eduardo's double-barrel, and were off.

The hike proved considerably more difficult than either of us expected. The slope down to Alder Creek was steeper than it looked from the ridge, at times exceeding forty-five degrees on loose rock with only a sparse cover of black spruce krummholz and little else to hang onto. It was pika heaven, with no shortage of the industrious little animals, and we descended to constant alarm squeaks. The climb back up the other side was equally steep and difficult, but once we were up, we found the site to be everything I had hoped for. Eduardo set about looking for beetles and I went right to work, laying out my ten-by-ten plots along a transect from the bottom of the slope to the top. I started my systematic search for tree seedlings and cored the mature trees that fell within the plots.

The days on that ridge were calm and productive. We were camped well above treeline, where we enjoyed just enough wind to keep mosquitoes from being a bother, but not so much as to keep insects from flying into Eduardo's malaise trap. It was noticeably cooler now, in part because of our elevation, but also because summer in the high country was rapidly waning. The dwarf birches were showing a hint of gold and

Eduardo with Malaise trap (see p. 45) in the alpine tundra.

the potentilla were turning that dark reddish-green that precedes its fall crimson. The bistort fruits were mature and ptarmigan were filling their crops daily, and the pika haypiles were full and curing in the pale sun, now nearly a month-and-a-half beyond the summer solstice. Wool replaced our cotton river-shirts in the mornings, and a down vest and hat felt good in the evenings.

Our daily routine was enjoyable, too. With only the three of us, our mission had become much simpler logistically. No longer did we have to manage boat, motors, gasoline, and individual schedules. Our interests were in one place, one ridgetop, one tent, and we came and went as we needed, on foot. We had food, too—lots of it. On top of our usual provisions, Frank and Mary Warren insisted that we take several thick salmon steaks from their freezer, and Eduardo was having considerable success bagging ptarmigan with his double-barrel. Maybe it was just the time and place, but the ptarmigan were surprisingly plump and tasted as good as any bird I'd ever had; or maybe it was the Oly beer that we washed it down with.

Garrett seemed particularly cheerful. It was no wonder that he was glad to be back in the company of others after his ordeal alone, but apparently it was the selectiveness of the company that made a big difference to him as well. Unbeknownst to me, Garrett had developed a marked distaste for Bruce's company while on the river, and for most of the summer had gone out of his way to avoid being in close quarters with him. Either Garrett hid his animosity extremely well, or I was so attentive to the personality conflict between Eduardo and Mark that it had completely escaped my notice. When it came Garrett's turn on the Yukon to make the milk run into Circle with Bruce, he passed, giving his seat to Eduardo instead. I recalled my amazement at the time that Garrett would turn down a chance at a hot shower and cold beer, but attributed it to his experienced comfort in the field. His comments took me by surprise now, for I was never more than amused at Bruce's eccentricities (did I not have a few of my own?), feeling that he added interesting diversity to the group. Nothing serious, though, Garrett assured me, he just couldn't deal with Bruce's personality. Nonetheless, Garrett was clearly feeling good now. Eduardo was also noticeably relaxed without the constant strain that existed between him and Mark.

For the first time since coming into the bush, I had only my own work to think about. Each morning, Eduardo and Garrett would putter around with a few camp chores and then go off in their own directions in search of unknown insects and small mammals. I would clean up a bit, pack some food and sampling equipment into a rucksack, and with the .444 in hand, head out toward my study site.

It was a long hike but I varied my course, found much of interest, and looked forward to the trek every day. I felt good, strong again, invigorated by the exercise and cool air. The rifle was heavy and I wished I could leave it at camp, but it was a sensible idea to carry it. When I went about my work, I moved slowly and quietly, often remaining still for more than half-an-hour at a time—long enough for a bear to wander into my study area without either of us knowing it. Others in a similar situation might sing, or shout, or blow a whistle to warn off bears, but I hated making noise deliberately, disturbing the great peace.

There were grizzlies in the area, too. At first I paid little attention to the overturned rocks that I encountered now and again, but as I found more and more of them I realized that this was the work of bears looking for ants. It was curious to me that an animal as large as a grizzly could find anything satisfying about a tongue-full of biting ants under a rock the size of a seat-cushion, but the bears were persistent in their search for food and left little uninvestigated. In time I began to understand that this foraging activity had implications beyond satisfying the needs of hungry bears: it also satisfied the germination requirements of a number of plants. In my search for tree seedlings, trying to understand the demographics of this treeline situation, I was finding very few, and with good reason. In the thick mat of tundra vegetation, there was almost no exposed mineral soil on which a tree seed could germinate and establish itself. In the end, the only suitable seedbeds I found were the exposed mineral soils beneath the rocks overturned by bears. So the bears, by their foraging activity, seemed to be contributing to the maintenance of plant populations that required periodic disturbance to the soil.

There was another connection between grizzly bears and conifer seeds in the northern Rockies that I had found interesting. This one involved an intermediary, the red squirrel. Where whitebark pine is prevalent, grizzly bears fattening for hibernation will eat almost nothing but the seeds of this tree during years of good cone crops. The bulk of these seeds, more than ninety percent of them, the bear will obtain by raiding the food caches of red squirrels.[11] Red squirrels are obligate hoarders of conifer seeds, an essential element to their winter survival, and will bring together the cones of conifers—thousands of them— into one great larder. It is pure paydirt for a bear to find such larders in whitebark pine country, but not so here. Red squirrels were actively caching spruce cones—upwards of ten thousand of them to get through the winter—but spruce cones are more woody scales than seeds, and the seeds themselves are small and contain only a fraction of the energy of whitebark pine seeds.[12] So, grizzlies in this country were bothering neither the squirrels nor me right now.

Nonetheless, I kept an eye out for bears, all the while hoping I might spot wandering caribou. On one return from my study site, I found an old caribou antler that served as a reminder that we were camped in the middle of Fortymile caribou territory. In times past, we might have enjoyed the spectacle of numerous bands foraging within sight of our tent, as caribou sought out the high country during summer for relief from mosquitoes. So plagued are these animals by mosquitoes that on an annual basis, female caribou with calves experience their greatest daily energy expenditure during the peak of the insect season; far greater than the amount of energy required to keep warm during winter.[13] But given their low numbers at present, it was hard to predict where the highly nomadic caribou might be now. Caribou are specialist ruminants that subsist mostly on a diet of lichens, and lichens are plentiful almost anywhere in this country. But the fundamental nature of this food source may also contribute to the far-roaming habits of the caribou. While locally abundant, lichens are slow growing and easy to deplete, so the ingrained nomadism of caribou may have evolved partly as a means of avoiding overgrazing in large herds. There was another force at play as well, one that worked against our encountering the caribou in any number. With insect harassment lessening a little every day, the caribou would be less inclined to bunch together. As the season turned, they would start to disperse, slowly moving toward the lower forest where they would again gather for the fall rut and spend the winter.

Garrett displayed his usual enthusiasm when I presented him with the rodent-gnawed antler. He was excited, it seemed, by all things mammal. But Garrett was having a particularly good day himself and had a counter-surprise for me: his second singing vole. He was filling a gap in the distribution map of this species now, having caught the only two specimens of *Microtus miurus muriei* so far known from the Yukon-Charley region.

We had been on the ridge a week and our last full day of work was at hand. It was a gray, overcast day, calm and not particularly threatening. I hiked over to my study site for the last time to finish up some note-taking and pull all of my plot markers. Any remaining evidence of my presence, I trusted, would soon be absorbed by the mountain.

I was glad to be working alone this day. Sometime after breakfast, maybe as I was packing my gear, I started to turn inward, to become introspective. For almost the whole walk over to my site, I was barely

Spruce clones and saplings at treeline.

aware of my steps, conscious only that I was on a mountain and the mountain at the moment was very calm. When I reached my site I methodically began pulling up my corner stakes, starting at the bottom and working my way uphill, being sure not to miss anything. When I reached the top of my transect, I stared back for a long time, partly to see that I had left no trace, and partly just to see. I sat down with my notebook and recorded a few thoughts.

I was still struggling over the initial implications of my work. White spruce was dominating the treeline here, as it was on Twin Mountain, and still forming clonal islands contrary to the current dogma. But the tree-ring record presented me with an even more difficult challenge. The increment cores that I was pulling out of these trees were telling a story that was equally untenable given the wisdom of the day. The mature founders of my newly discovered clones were showing a gradual, but sustained increase in annual growth increment contrary to normal growth trends. If a tree were somehow able to add the same volume of wood year after year, each growth ring would be slightly narrower than the previous years' because the tree is slowly increasing in both height and diameter. Generally, however, as trees age, more and more nutrients are sequestered in the long-lived wood, leaving the soil depleted and eventually stressing the tree so that it produces less wood over time. And some of these trees at the forest limit had been rooted here for more than two hundred years, longer than the oldest and biggest trees on the Nation River. Besides producing rings of gradually increasing width over the past several decades, these trees were reproducing successfully, and a quick look around told me that seedlings were establishing well above the present forest limit. All this, coupled with the general lack of tree mortality in the area, suggested that growing conditions were steadily improving at treeline—that treeline was advancing—and that was entirely perplexing. I was looking at two possible interpretations:

One was that the climate of this area was gradually changing, such that growing conditions were more favorable now than they were in the past. Scientists were talking more about the potential warming effects of increasing CO_2 in the atmosphere—the Mauna Loa Observatory in Hawaii had been documenting a steady rise in atmospheric CO_2 since 1958, but the climatic record from Fairbanks was not yet long enough to show any discernable temperature trend. Nor had anyone yet documented in the field plant response to rising CO_2 by itself. If the growth rings were reflecting climate effects in this area, they were telling a newsworthy story, but it might take another thirty years for the climatic record to become clearer.[14] Perhaps, I hoped, the age distribution of trees at the forest edge would also tell something about the favorability of conditions in more recent times for seedling establishment.

Failing that, my alternate hypothesis was that these trees were benefiting from a sharing of resources, that the whole was somehow greater than the parts idea—something that was not known to happen as a result of the cloning that was not supposed to happen with this species. Either way, I was looking at a challenge again. Out here in the middle of nowhere, where no one could look over my shoulder, I was tampering with the truth and the scientific community would not accept it matter-of-factly.

This is what is both difficult and rewarding about science. It is not easy to change our sacred views, but it is somehow satisfying when we do to know that we weren't entirely right, and that there is still plenty of room to grow in our understanding of our universe. I had thought about this a lot in the past few days. Truth is a chameleon. It seems so much a product of individual perspective, and perspective so much a result of individual experience, that I wondered if there could be any such thing as universal truth. Science trains us to be observers, to measure carefully, to formulate alternate hypotheses about what we see, and to draw conclusions based on the data. But science also teaches us what to expect, and therefore what to see. If a dozen observers, all looking at the same object from the same vantage point see the same thing, then that becomes the truth. If another observer goes around to the other side and sees the object differently, the truth changes, but only in the mind of the last observer. We measure, accumulate numbers, and let experience fill in the blank spaces. Measuring fifty below zero with a thermometer gives me quantitative information, a low point on a graph of cumulative data, but experiencing fifty below zero gives me an understanding that is entirely lacking in the number, and that shapes my truth. So truth must always be cited; the truth as according to whom.

Before finishing my graduate work, I witnessed a revolution in geology that overturned much of the science. Geologists had for more than a century been captivated by the notion that the shorelines of earth's continents fit together remarkably well if the continents were simply moved around a bit, suggesting they may once have been part of a great, single landmass that had subsequently broken up and drifted apart. Any notion of drifting continents, however, had been put to rest once and for all by the mathematical calculations of Sir Harold Jeffreys, an English physicist who convinced the scientific world in 1925 that continental drift was physically impossible. And that was that—the truth. Jeffreys had so thoroughly rebuked any thoughts about the possibility that it would be another quarter century before some dolt, a slow learner who obviously hadn't read Jeffreys, would mutter something again about continental drift and go poking around in the geological dustbins for evidence. By then, new discoveries were coming so fast that before I was halfway through my studies, every geology textbook still in print had to be rewritten for the proliferation of details being documented about seafloor spreading and the drifting of crustal plates.

Witnessing this revolution firsthand made me more than a little wary of absolute truths. It was not an unfamiliar feeling, however, as I had been there before, challenging church doctrine, and gotten into plenty of trouble for it. I was raised Catholic and for my first seventeen years was quite afraid to question the church's authority. Quietly, though, I was growing unsatisfied with the truths I'd been taught; unsatisfied with the notion of inserting a supreme being, especially one with a human likeness, as the answer to all the unanswered questions swirling in my head.

Perhaps this was the beginning of my leanings toward science; not so much for answers as for the process of sorting out the possibilities. At first my science was simply a fascination with the natural world around me and my questions were mostly about the identity and habits of things wild. But science really began to appeal when I realized that it was more about thought processes than collections of specimens or facts. Once again, I got caught up in the spirit of asking questions, rather than simply accepting answers, and that has proven infinitely more satisfying. Bringing this spirit to the consideration of religion, however, is probably the downfall of any orthodox practitioner.

Cutting myself loose from the anchor of organized religion, and all privileges with the almighty attached thereto, had its practical disadvantages. It would have been comforting when lightning was threatening to blast my remains four ways to the wind to believe that some supernatural being was protecting me (theists have assured me this was the case), but this idea is far too egocentric for my tastes. I couldn't be so presumptuous as to believe a lordly benefactor was looking favorably upon me when I share this planet with seven billion others, many of whom could use far more attention. I prefer instead to see myself as simply another species on the landscape, and the fact that I am a sentient, thinking being is either a blessing or a curse depending on how I am feeling at the moment, but makes no difference otherwise. When lightning was crashing around me on that ridge top, I was just plain scared out of my wits, exhausted from adversity, and hoping against hope (prayer not being an option for me) that the charged fields generating hundred-million-volt sparks between ground and cloud were not dancing around my tent.

Maybe this is all too simple. I have settled nonetheless into a position of comfort with my place in nature as part of a larger whole, somewhat akin to the Buddhist notion (with no further claim to Buddhist following) that nothing exists independently, without connection to all other things in the universe. This is only a natural outcome of my science, I suppose. This is what I study and I have not yet found any reason to remove myself from the ecological considerations of all the other organisms that I observe. I rather like Francis H. Cook's perspective in The Jewel Net of Indra,[15] in which he writes, "the most ingenious attempts of Western thinkers to erect a satisfying picture of existence has resulted in the not too surprising conclusion that while we are less than gods, we stand just below the angels, superior to and apart from all other things. One may ask whether this conclusion has not risen out of a pathetic self-deception."

What does set us apart from the other organisms with whom we share this planet are two things: our tool-using ability and our communication skills. We are not unique in possessing these skills, just better at it than most. Our use of tools, though not unknown in the animal kingdom, has enabled us to manipulate our environment in unprecedented ways, to create machinery and weapons by which we can compensate for other weaknesses, and subdue everything else on this planet. Our adeptness at communication allows us to pool information at a rate greatly exceeding that which any one individual could accumulate in a lifetime. Our communication skills also allow us to exercise dominion over those whose skills are less practiced, leading to the formation of elite classes that are able to exploit others for personal gain. Withholding information from others also confers great competitive advantages—starving out the enemy, as it were—and misinformation can be used as effectively as some of our most frightening weapons, to pit one faction against another, one religion against another, to deceive whole nations into acts of intraspecific annihilation unknown among other species.

This hardly seems a step below angels. If we are a lordly creation, I might ask what the Lord had against us. The way we're headed, our superior tool-making and communication skills may confer upon us the shortest reign of any species yet to appear in the geologic record. Evolution has its setbacks, too (". . . and the meek shall inherit the earth"). I can only hope that in the end, my remains will not be interred out of the reach of microbes that will suck up the atoms of my body and pass them on to other life forms, or free them to drift through the universe until they combine again into some other entity.

It started to drizzle. I pulled on my raingear and couched into a depression in the soft moss. I retracted a little, snugged my collar closer around my neck, tugged at my hat. I stuffed my notebook into my pack and set it behind me, leaning against it to keep it dry. I pulled my hands into my coat sleeves and crossed my arms. I was warm and comfortable.

Next to me was a young spruce tree only two-feet tall, growing alone, straight, a sapling that had established there on its own and was making it. Beside the sapling was a smallish, flat rock partly obscured by the low branches and delicate foliage of alpine bilberry. Perhaps, I thought, a grizzly wandered through here a decade or so ago, maybe when I was just returning to college for a second try, turning over stones in hopes of finding something to eat. Maybe a seed released from a cone on a nearby spruce floated down to this spot and found a safe site among the dense groundcover of sphagnum moss, bunchberry, and the trailing mountain cranberry.

Studying the young spruce in the drizzle, I watched moisture coalesce into a droplet caught at the base of a needle. Soon it was joined by another running down the stem. The two converged into one, hesitated there for a moment, then careened down the sloping branch until it ran up against a stout twig. As its momentum caught up with it, the droplet fell. Netted by a bunchberry flower below, it was sucked into the drying sepals then squeezed out by another drop to finally hit the ground. It would not be siphoned immediately into the tree root, though. Too many other droplets had soaked the soil ahead of it. Instead, it passed down through the soil to join others in a rivulet, then into the creek, and from there rushed down the mountain to the big river. Swept up in the current now with so many others, the droplet would drift to the sea where it would once again slow and wait its turn, wait to be plucked back into the clouds. If it played the currents right, perhaps it would show up next in the Peruvian Andes or the waters of Shanghai. Maybe in another million years—sooner if it chanced to fall in the desert, where its residence time would be shorter—these same molecules might pass this way again.

Then the drizzle let up.

Gordon came for us earlier than expected the next day, and he brought with him a change of plans. It was time for us to return to our base camp on the Kandik and finish our field season exploring the lower Charley River. Bruce should have returned by now from his fossil hunting further up the Yukon and was to leave the boat in Circle. Steve Young was due to arrive any day from his Leningrad trip and would take the boat, with supplies, upriver to find us. Gordon was to fly us directly to the Kandik and we would busy ourselves back on Kathul Mountain until Steve showed up.

Things had gone smoothly for both Bruce and Steve. The boat was in Circle as planned and Bruce was on his way home, and Steve had just arrived on the mail plane that morning. But it took no time at all for Steve to hear the story of our Charley River incident and now he wanted to know more. So, he asked Gordon to bring me directly back to Circle, instead of to the Kandik. Gordon would then take Garrett and Eduardo to our base camp and I would accompany Steve up the river.

But Steve wasn't the only one who wanted to know more when I got back to Circle. There was another helicopter, the second Alouette I had seen in a month, parked at the end of the runway when Gordon and I returned. We set down beside it and waited for a minute as the rotor wound down.

Gordon took off his headgear. "Looks like we have company."

"Yeah," I said unenthusiastically. "I'm not sure I'm ready for it. That looks an awful lot like the last Alouette I saw out here."

Steve came out to greet us. He was trailed tentatively by a neatly-dressed stranger, tall and straight-postured, trimmed hair graying at the temples, clearly not a woodsman. Steve wore his usual disheveled look, a broad, almost childish grin, happy to be back on Alaskan soil, happy to reconnect with the expedition he had conceived three years earlier. And for the moment, at least, I was that expedition. Steve was happiest in the field and after a long, roundabout flight from Leningrad, was ready for some river time. His giddy handshake spoke wordlessly of all these things. Then Steve turned to the tall stranger standing awkwardly in the background and introduced him. Chief Pilot, Anchorage Helicopter Service. In civilian aviation hierarchy, Chief Pilot is top commander. By coincidence, he and Steve had arrived in Circle at the same time, and when the stranger went into the trading post and asked if anyone knew where he could find me, it caught Steve's attention.

The man was serious. Everything about him was serious—his demeanor, his speech, his dress—and so were the charges he had come to investigate. Leaving people in the bush, under duress and without gear, without the emergency drop box all bush pilots are supposed to carry, and then not reporting it, was serious. Criminal-negligence serious. This is what Steve walked into, and now he too wanted to hear it all. So the three of us went into the post, leaving Gordon to finish airlifting Eduardo and Garrett to our base camp. Over a beer and burgers, I told my story. The stranger scribbled notes as fast as he could. When I stopped, the questions came.

In the comfort of the trading post, with a couple of weeks of satisfying field work in the high country behind me, it was easy to question our decision to stay put on the Charley. It was a decision, though, that was based on expectations that someone would come looking for us, not that walking out would be impossible. Difficult, yes, but only a question of time if we were reasonably careful. Unlike others, though, who had traipsed all over this country in the course of seasons or years, we felt an obligation to finish a job, and walking out the Charley would have greatly complicated matters of communication. When someone did come looking for us, where would they start? Where would they go next? When we made it to the Yukon, how would we get word to Garrett, or Gordon, or anyone as to where we were. We'd either have to get down to Circle by ourselves or catch someone on the river going that way and trust them to deliver a message for us. And then wait somewhere else, maybe with scrounged food, but to no other advantage. No, to walk out the Charley might extend our downtime considerably. But the problem with our delayed decision to raft out was that we had handicapped ourselves with seven days of fasting before starting.

The stranger had little to say but that he would take corrective action and it wouldn't be necessary for us to file a report with the FAA. "And if I can provide you with whatever additional air support you might need to finish the job, as compensation for the time you lost, I'd be glad to." Anything, he might have added, to keep this from getting to the FAA.

Then he asked, "Mind if I go out and talk to. . . which one was dropped off alone. . . Garrett?"

"I think you should. Garrett's the only one that knew what was on Mike's mind when he left us."

"Would I be able to find him easily?"

"Gordon should have 'em back at the Kandik soon. They'll be in plain sight at the mouth of the river, on a wide bar with plenty of room to land. Look for their big canvas tent."

He folded his wire-bound notebook and got up from the table. "I'm really sorry about all this. You've got my direct line now. . . just let me know what I can do to make things right."

Steve and I stood up, shook the stranger's hand, and watched him walk out to his helicopter. Then we sat back down. Steve just gave me a look that said "can't believe it."

"I'm surprised you went right back out after Gordon found you."

"Yeah, well Eduardo wanted to hang out here for a couple of days, catch up on some letter writing and do some laundry, but I'd already done my letter writing. Gave 'em to Mike to mail when he got back to wherever the hell he was in such a hurry to be, but God knows if he ever did. Anyway, I had no interest in staying here and was chompin' at the bit to get into the mountains and get something done. And Garrett was game. He didn't do much up there but try to find us and keep his own sanity, so he was fine with going back and setting some traps, long as he had some company. And Mark? Well, I wasn't going to subject him to any more. He'd had enough and really needed to be on his way home, so everybody was happy."

"You ready to go back upriver now?"

"Quick as we can get that barge loaded. And I'm really eager to hear about your trip."

"Yeh, we'll have lots to talk about, but I'd sure rather do it around the fire."

It was early evening when we nosed our boat into the current and plowed slowly out to the main channel. Once again we were loaded to the gunnels, carrying upwards of 150 gallons of gas—enough to get us as far as the Nation again, plus a good distance up the Charley this time—and food enough for another week. Still, with only two of us and considerably less gear than we carried on my first trip up the river, I figured we were five hundred pounds lighter and would probably make base camp in eight hours or so. What I didn't count on was how much more challenging river navigation would be by midnight. Day length had shortened nearly three-and-a-half hours already from my first run up the river, some seven minutes a day, enough to make a difference in night travel. We reached the Kandik at 2:00 am and roused Eduardo and Garrett out of their sleeping bags.

We lazed around camp the next day, Steve talking about his Leningrad trip and napping off his jet lag, the rest of us catching up with camp chores, writing in our field journals, washing clothes, bathing in the cold Kandik. Eduardo walked up the slough with his fishing gear and came back with two pike. Fresh fish, old potatoes, and canned beans for dinner that evening.

The days that followed were relaxed and comfortable. Steve brought renewed energy to the group, fresh enthusiasm for the place and what we were doing there. It felt no longer a job, though. We were simply living, day to day, with the fullest awareness of our surroundings. We felt at home, settled into a rhythm as natural as the river flowing. The nature of our conversation was different, seasoned, matured. Behind us were our initial anxieties in Circle, constant attention to logistics and support, seeing that everyone got what they needed. Bruce, who had been all about academics, and Mark, who lived entirely in his head, were out and the four of us felt the ease that comes with familiarity, acceptance, self-confidence.

Eager to return to the slopes of Kathul Mountain, Steve and I off-loaded our extra gasoline after a day of rest and took the boat upriver while Eduardo and Garrett busied themselves setting out their traps for a second round of collecting near the campsite. It was fun to see Kathul again, this time through Steve's eyes. It was easier going, too. We had already worn some semblance of a trail through the half mile of dense willows and alders near the river, the air was cooler now, and the mosquitoes were down considerably.

A short way up the slope, Steve stopped amid several clumps of *Podistera yukonensis* and looked around pensively. Like Bruce's reconstructing ancient landscapes from his fossil specimens, Steve seemed to be looking through a portal into the past. The inconspicuous species at his feet—a small tufted plant not four inches tall with blue-green leaflets and a creamy-white flower cluster much like that of carrot or parsley—was among the rarest of Alaska plants, previously known from only three other locations in a state one-fifth the size of the Lower Forty-Eight. Yet it was growing all over the slopes of Kathul. Steve had discovered the plant here the previous summer and it's partly what got him to thinking about the possibly relic nature of this whole hillside. That and the prominence of sagebrush here. Sage was a part of the puzzle as well, and not just for its abundance. Of the three species of sagebrush on this dry slope, one, *Artemisia frigida*, was fairly common throughout the Rocky Mountains. The other two were a different matter.

"Look at this," Steve said as he picked some leaves off a low sage near the *Podistera*, automatically crushing and sniffing them. "This is *Artemisia alaskana*. It's supposedly endemic to interior Alaska and the Yukon Territory, but Boris Yurtsev told me in Leningrad that he thinks it's the same species they are calling *Artemisia kruhsiana* in Russia. It grows in situations just like this throughout Siberia and you can't tell 'em apart in the field except by which side of the Bering Strait you're standing on. Turns out they're even similar in chromosome number."

He looks around from where he is standing. "And *Artemisia laciniata*," he adds, pointing to another low shrub, "the third species of sage here. The only other place it's been found in the New World is over in the Tanana Valley, but it's all over central Asia. Seems pretty likely its distribution was continuous when the land bridge was exposed."

Steve put his pack down.

"Mind if I spend some time looking around?" He didn't need to ask.

"Be my guest," I said. "If I knew how to help I would, but I'm real happy to sit here and enjoy the place. Last time we were here it was hotter'n hell and the mosquitoes were outrageous. We didn't take much time to enjoy the scenery."

Steve broke into his giddy smile again. He was back in his element. An expert on the vegetation of the Beringean region—he had spent his years as a Harvard doctoral student, studying the flora of Saint Lawrence Island off the coast of Siberia—in one afternoon, Steve found four additional plant species that would prove equally rare and endemic to the Kathul area. The association of this many endemic species in the same place hinted strongly, in Steve's mind, of a relic plant community long cut off from a larger world. Excited as he was at the moment, he could not have imagined how much attention this would garner in the coming years. For the next two decades, others would detail every aspect of this hillside, its climate, vegetation, and soil characteristics, and debate its authenticity and importance as a living remnant of a biome that, ten thousand years earlier, may have extended from here to northern Europe.[16]

The next day, the four of us headed downriver for the mouth of the Charley. Eduardo and I had a pretty good feel for the upper canyon, having descended one of its tributaries and walked a few miles up and down the river looking for Garrett before settling into one spot for those discomposing seven days of intensive observation, but we had made only a brief incursion beyond the mouth of the river earlier in the summer. So we all looked forward now to exploring its lower reaches in the boat, Garrett especially, since he had missed the upper Charley experience altogether.

The first hints of autumn in the high country had by now descended to the river. Signs were subtle, but there if you knew where to look. The shortening days were making an impression on plants and animals, though not so much in outward appearance yet. Tracking seasonal changes in day length was the primary means by which plants and animals timed their advance preparations for the cold ahead. And cold was coming. The previous summer's growing season in Eagle, the length of time between the last frost in June and first frost in August, was officially seventy-seven days if you don't count the freeze recorded by the weather observer on July second. Long days around the summer solstice can make up in part for a short season, but nothing compensates for freezing. If a plant is to withstand freezing without injury it must shift into acclimation mode in a timely manner, shutting down certain metabolic processes and starting up others, preparing cells for the stresses of winter. And timely meant now. So the green shades of summer were a little lighter, more yellow-green, as plants were beginning to conserve nitrogen and phosphorous; not actively recycling it yet as they soon would, but no longer rebuilding chlorophyll as it broke down, thus leaving the greens a little thinner in color.[17] Whatever the greens lacked in intensity, however, the fireweed made up for. In full bloom now, the abundant, rose-pink flowers of fireweed panicles brushed swaths of color across the landscape wherever local disturbance, especially fire, facilitated their spread. Fireweed is visual compensation for blackened spruce in the boreal forest. Animals, too, seemed to be anticipating the coming season. The sea ducks that moved inland to rear their young in the countless bog ponds along the Yukon flood plain—the scoters, buffleheads, goldeneyes, mergansers—were now gathering in increasing numbers at the mouth of the Charley, soon to migrate in large flocks back to coastal wintering areas.

Turning up the Charley felt like exiting a busy highway onto a quiet back road. Not that there was any traffic on the big river that morning. The Charley, running smoothly and darkly transparent in its lower reaches, imparted a feeling of calmness, almost relief, as we came off the grey, silt-laden Yukon. It was as if we had just shaken some invisible behemoth bearing down on us from behind, sidestepped safely out of its way into still waters. Even the Charley seemed intimidated as it emptied into the Yukon, its clear, cold water hugging the left bank for a considerable distance downstream before finally mixing with the big river. We trimmed back on the throttle until we were barely making headway, and relaxed.

The rushing, white-water, boulder-strewn Charley of the upper canyon turns languid toward the Yukon, its looping bends sometimes almost doubling back on themselves. From high altitude, the thin ribbon of the lower Charley looks like a tangle of old fishing line stomped into the moss. It curls and loops, splits and rejoins, ties itself in knots as it meanders across the flat terrain, weaving together the visible remnants of old river bends cut off from the main channel and slowly reclaimed by bog and forest. In the endless cycle of river migration and plant succession, the landscape is constantly shifting in its parts, but the whole remains the same: a complex mosaic of oxbow ponds, muskeg, river bars, and spruce forest in various stages of development.

On a heavily wooded stretch of river, Garrett spotted a covey of spruce grouse gathered on a narrow, gravelly beach. It was a rare sighting,

River meander patterns on the lower Charley River.

though I wondered why. For all the white spruce growing along these rivers—the bird is intimately tied to its namesake tree for its winter success—the spruce grouse had proven particularly evasive. The bird is distributed widely across the entire North American boreal forest, but it is reclusive, well camouflaged, and has a habit of holding still in the face of danger rather than taking flight.

Superbly adapted to life year-round in these forests, the spruce grouse has found a way to tap an unlimited source of food when all else is under snow. It subsists during the winter almost entirely on a diet of spruce needles. It's not an easy trick, though. Spruce needles are more cellulose and lignin than anything else, and are loaded with defense compounds designed specifically to discourage browsing. Extracting anything of nutritional value from a diet like this requires more than ordinary digestive abilities, which explains their gathering on that gravely river bank.

Ames' cabin at the mouth of the Charley River.

As the grouse shifts from a diverse summer diet that includes tender greens, insects, fungi, and berries, to one of tough spruce needles, the digestive tract of the bird undergoes a remarkable change. The crop, gizzard, large intestine, and caecum all increase dramatically in size. In conjunction with this, the birds at this time of year will seek out and ingest grit—sand or gravel—to improve the grinding of coarse, fibrous needles in the gizzard.[18] Stream banks or uprooted trees are the most common, often the only sources of this material, and experienced grouse will make sometimes lengthy overland excursions to familiar sources of grit, where they may congregate with others for several days before dispersing to their winter territories.

We stopped to check out the old Ames cabin where Leon Crane ended his long walk, then a few more miles upriver we beached the boat to have a look around. My first step onto the silty shoreline landed beside a wolf track as wide as my boot and five-inches long. The footprints of the lone wolf mixed on that narrow strand with moose and lynx tracks, each coming and going on their own schedules, hunter and hunted overlapping territory in a time-honored cat-and-mouse game, ending mostly with the same winners and losers.

We had been told there were four wolf packs on the middle Charley and according to Frank Warren, who flies over the territory often, wolf numbers had been up for the last few years and the moose population down. This arithmetic made sense on the surface: when wolves are up, they may be the most important factor controlling moose numbers. But predator and prey in this case don't always track as closely as with the lynx and snowshoe hare. Other factors can put the two out of sync. When wolf numbers are down, climate and forage availability become important regulatory factors for moose. Moose are superbly adapted to life in the cold, but exceptionally deep snow may increase energy expenditure and mortality among calves and ageing moose in particular. And ironically, unusually warm summers may also increase winter mortality as moose experience heat stress at temperatures approaching seventy degrees, suffering increased metabolism, loss of weight, and failure to accumulate sufficient fat reserves critical for winter survival.[19] Moose also depend heavily on early successional plants like willows, birch, and aspen, which means that periodic disturbance to the forest, such as fire and flooding, is essential to their long-term stability. So the fortunes of moose may rise and fall with a number of environmental factors, but always they are under pressure from the top. When the ratio of wolves to moose is high, as apparently it was now, the arithmetic does not favor the moose. Until the wolves experience a decline, the moose will likely remain down. It is an epic struggle between hunter and hunted.[20]

Out of the boat, Moose, the dog, wasted no time tangling with an Alaska-sized porcupine—another time-honored contest mostly with the same winners and losers. While I held the whimpering dog, Eduardo yanked two dozen quills from its nose.

Fred pulled into our camp shortly after we returned. He was bare-armed again, wearing his olive T-shirt with the floppy brown hat, same as when we last saw him on the river. I met him at the water's edge and grabbed the bow seat of the canoe to haul it onto the sand.

"How's it goin' Fred? Been a while." I looked into the canoe. "No beaver? I was looking forward to another feast."

"I 'magine you was," he said with a smile.

"Yeah? You think we're starving out here?"

"Hear you guys had a little trouble up the Charley."

I stopped short and looked straight at him. "You know about that already? Jesus, Fred, for all the trouble we had trying to communicate up there, word sure seems to get up and down this river easy enough."

"People on the river are watchin' you, curious to know what you guys are doin'. When somethin' like that happens, it gets around."

"Were they cheering or what?"

"Just wanna know what's goin' on, that's all. So what'd you do up there?"

"Not much. That was part of the problem."

"I mean what'd you do for food?"

"Answer's the same. Not much to eat up there when you haven't got a gun. No luck trapping fish. Blueberries were green. Stoned a couple of red squirrels, but they don't amount to much. And hell, Ed here's supposed to know everything about bark beetles but couldn't find us a single damn grub to eat when we needed it." I grinned at Eduardo.

Fred followed me up the river bar and plunked himself down straddling the big log in front of our cook tent. "Wish I'd a been with you."

I pulled a bucket out from under our makeshift table, flipped it upside down and sat on it. "So do I, Fred. I might have learned something."

I told him the whole story and then turned the questioning around.

"So what would you have done if you were up there?"

He shrugs. "Guess I'd a set some deadfalls. No shortage of heavy logs up there I 'magine. Or big flat rocks. Rabbits are down now but that don't mean there ain't any around."

"Hares, Fred."

"I'd use those squirrels to bait somethin' bigger. There ain't nothing in them to eat anyway. An' rabbits'll come to meat. They ain't jus' vegetarian."

"Yeah, I heard that once from an old trapper in Vermont. Was never sure whether to believe him or not."

"Take rab. . . hare, boil everything up together, makes a pretty good stew. Maybe throw in some of them green blueberries. Or fireweed. Ain't no part of fireweed you can't eat. Flower buds'r good right now."

"You make it sound like I had a kitchen up there, Fred. What the hell was I supposed to use for a pot?"

"Cut yourself a big piece of birch bark an' fold it into a pot, kinda square-like, an' be careful you don't put no holes in the corners. Soak it good first. Tie it together up top with whatever you got, spruce roots work good, an' hang it with a couple sticks."

"Birch bark! For Christ sake, Fred, that's what we start fires with. You're telling me now I'm gonna hang a birch bark basket over a fire with the makings of a stew in it and it's not gonna go up in flames!"

"Long as it's got water in it, it can't burn through."

That one stopped me cold. I had to think about it for a few minutes, letting the logic sink in. "I'll be damned," I muttered to myself. I gotta try that someday. Try baiting a snowshoe hare with meat, too.

It was all pretty much matter-of-fact to Fred. Getting dropped off nowhere would probably not have meant much to him, but then he traveled lighter than most. He offered a piece of moose jerky as we sat. We talked more about deadfalls and snares, how to make the best triggers, and what else we might have caught. Fred was savvy enough. He didn't trap as much as others—just did what was needed to get by, and no more. That seemed to be his way of living. But he was interested in his surroundings and absorbed as much as he could from our conversations. He asked questions, too; about the fossils we were finding and what they told us; why we were so interested in the sagebrush on Kathul Mountain; what kinds of small mammals Garrett was finding. As we talked, we'd periodically push the firewood, long pieces of driftwood burning at their ends, into to the center of the campfire, keeping the fire going for its warmth as cold air settled into the river bottom.

The conversation stalled. We were both sitting there, staring at the fire when Fred spoke again.

"You're takin' to this country, ain't you. You seem pretty easy, like you could make a go of it here."

I dwelled on the question for a minute, not quite sure how to answer it. Then I looked at him.

"You know, Fred, from the time I was old enough to read books about the North, I've dreamed of being here. I'm not disappointed." I let it go at that.

Later in the evening, when things had quieted down, Steve came over to where I was sitting. It was dark now, and chilly. In his quiet manner he stood in front of the fire for a few minutes, then turned his back to it, warming before he spoke.

"You get all the data you wanted this summer?"

"I got enough to keep me busy for a while."

He turned and poked at the fire with his foot, stirred some coals into flame, then looked at me again. "I'm going back down the river in the morning. Why don't you come out with me?"

Pause.

"Air North comes in tomorrow. We can probably get you and Garrett on the return flight. Get you started on your way back to the States."

I stared at the ground for a long minute, looked at my hands, the fire, back to my hands. Steve stood silent, waiting.

Finally, I answered. "I don't know, maybe it's a good idea. . . . I'm gettin' awfully comfortable here. Might be the last boat home for me."

VII. THE YEARS AFTER

DEATH WITHOUT REASON ✦ MOTHER OF ALL SUMMERS ✦
PLANTS ON THE MOVE ✦ CHANGING FORTUNES OF MOOSE
AND MEN

March 7, 1983. The day dawned with all the promise of spring in interior Alaska. The long nights of winter were over, the sun climbing ever-
higher above the southern horizon, adding more than six minutes to
every day, an hour more of sunshine every ten days. Only two weeks
earlier, temperatures were peaking midday at fifteen below zero, but
now the thermometer was pushing against the freezing mark. Winter's
hold on the landscape was loosening.

It was a quiet time in Circle City, a time for soaking up the new day-
light, thinking about gardening, making plans for a busy summer. The
waning snowpack—officially twenty-four inches deep—sparkled on the
surface from the night's hoar frost and the sun was warming fast. It was
the kind of day in the North where you look for something to do outside; shirt-sleeve weather to an Alaskan. Gordon decided it was a good
day to move the remainder of the winter's woodpile away from the back
door, back into the woodshed. He was living in Circle full-time now,
settled into a comfortable modular home with his wife Lynne and their
two young daughters.

Gordon stepped out the back door, stacked a load of firewood on his
arm and carried it to the old shed. The shed, built for hay storage in
the early 1900s, looked like a lot of out-buildings in Circle: weathered

gray and leaning a little to the south, the product of ceaseless snow blast, sun bleaching, freezing and thawing over the decades. Gordon stooped with his load and entered the shed where the cold, stagnant air of winter still hid. What he could not have known as he stepped into the shadowed interior was that the warming sun, radiating energy to the snow surface, was slowly settling the snow on the roof of a garage next to the shed. Minute forces, transmitted downward from one ice grain to the next in imperceptible steps, were upsetting a delicate balance within the aging snowpack. The loosely bonded crystals at the base felt the strain and suddenly collapsed, releasing the energy stored in the snowpack. Without a hint of warning, the heavy snow slid off the roof, adding to the snow load on the shed.

No one heard the shed cave in. The load of snow was sufficient to muffle the crack of stressed beams finally giving in to their burden. Gordon was too far inside the shed to escape. He caught the tonnage of collapsing wood and snow full force and was knocked lifeless. Found two hours later by Lynne, her brother-in-law, and father, Gordon was flown to Fairbanks and a waiting ambulance, but no amount of CPR on the way could bring him back. The two-tour veteran of enemy fire in Vietnam died while building his dream in Alaska, in a quiet village at the end of the road.

Many years have slipped by since our party explored the muskeg and forested hills of the Yukon-Charley wilderness. Eduardo went back to Seattle, married Pam, finished his degree, and eventually returned to Alaska for a career with the US Forest Service. Bruce, as expected, went on to make a name for himself in the ivied halls of academia. Mark took up a quiet life in New England. For Garrett, the years following were difficult. Having lost a tenure decision at the University of Rhode Island and ending up divorced to boot, he drifted around on the coast of Maine for a few years, doing field work for the state on some of the unpopulated islands offshore and holding a brief teaching spot at a small, private college before it went bankrupt. After an unbearable stint in a high school biology classroom, excitement returned to Garrett's life when he landed a job at Moi University in Kenya. To a mammalogist of Garrett's ilk, it must have seemed like reward in heaven: a continent with mammalian diversity like none other. But heaven was cut short. In May of 1987, the gentle biologist who had complained in his field journal of being short of breath when he was desperately alone and searching for his colleagues high above the Charley River, died of lung cancer.

I am writing now from a 10th Mountain Division Hut on a cold ridge just north of Mt. Zion in Colorado, only a raven's flight from the Continental Divide. At eleven-thousand-seven hundred feet in elevation, the air is snappy, the snow and the silence deep. In the thin night air, fine ice crystals seem to form spontaneously and lightly prick my cold cheeks and nose when I step outside. I have made a life of studying plants and animals in winter and this is my environment now. I have traded high latitude for high elevation, but much about the setting here reminds me of the Yukon-Charley uplands in interior Alaska. At something over eleven-thousand feet, the forest begins to break up. The trees still stand tall and close for the most part, affording themselves a degree of self-protection, but with increasing elevation, openings or glades become larger and the self-protecting forest shrinks until, like treeline in interior Alaska, it is finally reduced to a few isolated tree islands. Above me, the snow-covered tundra rises yet another thousand feet to a soft, rounded ridgeline. The white-tailed ptarmigan have retreated from their summer habitat in the tundra to the shelter of scattered trees and deep snow. In the willows near treeline, they have been feeding, filling their crops with buds and roosting in shallow excavations beneath the snow.

I have been on snowshoes in deep powder nearly all day and am fairly exhausted, but I have crossed many tracks and feel exhilarated for the effort. Lynx are staging a comeback after reintroduction and snowshoe hare are plentiful. In the spruce forest below, red squirrels are doing well, and so too appear the marten. Their tracks weave through the trees in fast, efficient lines until diverted by sight or smell of prey, most often a squirrel, whereupon the marten changes direction seemingly in midair. A faint drag-mark in the snow, where no part of the marten would leave one, tells me when the hunt has been successful and the marten is carrying prey to a den site. The tracks leave me thinking about the likes of Morris Gundrum and his half-brother Silas, and following them, Fred Beech, Dave Evans, and any number of others trapping on the Kandik River, the Charley, the Nation, people who would regularly travel thirty- or forty-mile circuits on their traplines, taking fifty or more marten in a winter.[1] Lynx, hare, marten, ptarmigan; I am tempted to draw comparisons with the upland summits of interior Alaska, but I am cautious. Conspicuously absent are the large mammals—moose, wolves, grizzlies, caribou—but there is another element missing, too. There are no woodsmen here. No one that lives by the gun, trap, and net, that must read the weather and know the seasons, must cut wood in summer or go cold in winter, must find prey year round or go hungry. The people I meet on the trail are comfortable, sporting the best and most fashionable gear on their way to a warm welcome at the end of the day. I am among them, but I understand the difference.

There's another reason to be cautious about comparisons. Much about the environment and culture on the big river has changed. Northern latitudes have borne the brunt of global warming in the last half of the 20th century, and the results have been profound. When so many years ago I pulled a few increment cores out of spruce trees growing at their altitudinal limit and saw steadily improving radial growth,[2] I couldn't have anticipated how much change was coming. Average annual temperature in interior Alaska has increased steadily over the past few decades, at twice the rate of warming to the south, but this statistic tells only part of the story. The greatest change has come during the winter and spring months, both warming upwards of 8° Fahrenheit since mid-twentieth century,[3] with nighttime lows in many places rising faster than daytime highs. These nuances can have biological consequences considerably out of proportion to the trends plotted on graph paper. Earlier snowmelt speeds warming of the soil and, when coupled with a small increase in nighttime lows, can stretch the growing season notably at both ends. Combined with even slight warming, a lengthening of the growing season by two or three weeks—a conservative possibility—is likely to be especially significant near the limits of tree growth, particularly if low temperature is the dominating constraint at treeline, as is often supposed. In any case, the result of climate warming is showing on the ground. Whatever causal factor is doing the driving, trees and shrubs at high latitudes and high elevations are on the move. A pan-arctic vegetation shift is presently underway, with willows, alder, dwarf birch advancing into the tundra throughout the North. Treeline, too, is advancing in many locations.[4]

I will not lay claim now to having seen this coming. I could not have known then how justified those early atmospheric scientists were in their concerns about rising CO_2 levels. I could not have been so sure that forest stands high up on Twin Mountain and the Woodchopper Creek headwaters were responding to something we couldn't yet see in our climatic data sets. Tree response to climatic variables is, after all, complex business[5] and that little chameleon named Truth can be mighty slippery. But rising CO_2 and its potential effects on plant growth and climate has been coming on, slowly and subtly, for a long time—and time usually makes for better science. Alaska has now become the poster child for climate change.

This is not to say whether any of this is good or bad, but it is change, and the spruce bark beetles, at least, think it's all good. With earlier spring and warming summers, especially since 1993—the "mother of all summers," by Eduardo's decree—bark beetles have managed to shorten their life cycle in the boreal forest from two years to one, frequently enough to generate huge population increases from time to time. The effect is dramatic, turning entire forested landscapes to a rusty red-brown color; over two million acres in the 1990s alone, and the heat is still rising. These outbreaks can be enough to overwhelm even healthy trees, but the problem is exacerbated by increasing drought stress in the white spruce forests.[6] Alaska's interior, blockaded by two great mountain ranges, gets no more precipitation annually than Tucson, Arizona, and as summers warm and push evaporation rates up, the big trees along the rivers feel the pinch. They garner fewer nutrients from the dry soil, produce fewer defense compounds, and fall easy prey to the bark beetles. The bark beetles, says a colleague of Eduardo, now have a forest management plan of their own and, unlike the Forest Service, are unencumbered by bureaucracy and have legions of loyal followers.

With or without legions of bark beetles, fires are eating up a lot more forest, too, as summers in the Interior become warmer and drier. For the Yukon-Charley area alone, the trend is ominous. In 1986, the Eureka Creek fire on the right bank of the Yukon burned 44,749 acres. In 1991 another 40,000-plus acres burned in the Charley River Drainage.[7] In 1999, four separate fires in the preserve consumed a total of 153,736 acres. Then came the mother of all fire seasons. In 2004, during an exceptionally hot and dry summer, more than 455,000 acres burned within and adjacent to the Yukon-Charley preserve.[8] Some of this acreage was just recovering from previous fires. While boreal forests are fire-adapted in many ways, climate change may be forcing a shorter fire return interval, which will keep forests in earlier successional stages of development.[9] Moose will thrive, caribou will not.

As the forests go, so go the people, though not always from the same forces. Most of the homesteaders are gone from the Yukon-Charley now, driven out not by age or hunger or loneliness, or any number of other reasons a person might finally pack it up and leave the country. Except for a few that stay to the end, there has always been a slow turnover of homesteaders along the river; people drifting down the Yukon, staying for a few years, then moving on. Old cabins eventually melt into the ground, new ones get built, and things pretty much stay the same. Like a pool in a stream, the current flows through and the water constantly changes, but the pool remains a pool. Now, however, the pool seems to be drying up. The land has been taken once again from the south, and the new landlord seemed bent from the start on ousting the former tenants.

When the Park Service came into the country in 1980, things changed quickly. The government's mission was to protect the natural and cultural history of the river, but it seems the administration cared a lot more for the natural than cultural; or maybe had a different idea of what constituted history, and the existing culture on the river didn't fit. So they drew a line on history, wrote rules that one by one drove the culture out, and now the Yukon-Charley, as Dan O'Neill so aptly summed up in his book title, is *A Land Gone Lonesome*.[10] Except for the very few who had legal title to a small plot of land, most of the long-timers are gone; ironically, many as far south as they could get, as if to neutralize their years of cold, dark winters. Dave Evans ended up in the Chiricahua Mountains on the Arizona-Mexico border, Brad Snow took up blue water sailing in southern oceans, and even Fred Beech made it all the way to Jamaica, where he married again and eventually died. And there is no longer a slow current of newcomers to take their place. The region has emptied of its self-sustaining culture through attrition and prohibition. Living off the land is no longer a permissible activity.

I had never made it to Eagle. The Nation River was as close as I got when I was last in this country, but now I wanted to see it. The Yukon-Charley Rivers National Preserve has been on the map for a few years and I was curious to see the park headquarters, see what the town looked like. On our way up the Taylor Highway, my son Greg and I stop at a bar in Chicken (ostensibly named for the ptarmigan that nobody could spell) to wash down the road dust with a cold beer. The bar is packed with young people gathering after a day of summer wage work to make time as fast as they can. The beer drinking is hard, the social interaction feels almost rushed. The gender ratio in the bar is strongly skewed toward the male, so the women don't lack for attention. For entertainment, four inebriated twenty-something guys roll out a huge tractor tire, one climbs into it, the others give it a good shove downhill, and off they run after it as it bounces and gathers speed until it veers off the gravel road and crashes into the brush. When it stops, the guy inside stumbles out, not much dizzier than he was to begin with, everybody laughs with great animation and hand slapping, and the four push the tire back up the hill for someone else's turn.

That night we find a spot to camp just off the Taylor Highway. This is Alaska. But the next morning we are awakened in our tent by a tour bus stopped along the road. It is an outrageous violation and we are peeved, but soon we are laughing ourselves silly as we plot revenge, talking about bursting out of our tent, naked, blasting the air with our shotgun as we loudly curse the mosquitos, imagining the tourists panicking to get back on their bus. The stories they would tell!

At a high point on the road we pull off the shoulder and get out to enjoy the expansive view. We are scanning it with binoculars when a car stops and the passenger rolls down his window. "See anything?" he asks. "No" is all I said. ("Just a million acres of gorgeous wild land encompassing half the Fortymile watershed, full of moose, bear, caribou, Dall's sheep, and wolves.") The car speeds down the gravel road. At the edge of Eagle we read a large sign with a brief history of the town carved in wood. It talks about Fort Egbert and the old days; says the road we just traveled was used by prospectors, freighters, and trappers; that some of their old cabins still stand and are in use today. It says the telegraph line and fort are gone now, but "memories still linger on in the minds of the sourdough." Sourdough? They still here? In Eagle we browse a new art gallery.

Any place on the map labeled 'National' is bound to be a magnet. It's usually a different color and it draws the eye like a light in the night. It suggests adventure with little risk. It invites us to visit. Especially when it is new. Especially when we can drive there. When the Yukon-Charley Preserve popped up on the map, it did not escape notice. On average, six thousand three hundred people visit the preserve every year now. In some years the number is twice that. Each of the summer months will see one thousand five hundred people check-in at headquarters, but in 2010, in the month of June alone, three thousand two hundred people descended on the river. They stay an average of five days.[11] The land is no longer lonesome.

But the new river-people are different from the displaced home-steaders; a transient, recreation culture passing through with res-ervations, permits, trip itineraries, maps, and GPS units to tell them exactly where they are and track where they've already been. The new river people will never know the size and emptiness of the country as Lt. Leon Crane knew it when his B-24 crashed at the headwaters of the Charley, his four compatriots killed, himself eighty days getting out;[12] as Garrett knew it when he could not communicate his fears; as countless others knew it and whose stories will no longer be told. Nei-ther will they know the different kind of loneliness which O'Neill talks about: the emptiness of a land devoid of humans living on it, prosper-ing through their knowledge of the forest, river, and its seasons of wild harvest, through their skills as hunters, cabin builders, dog mushers, boatsmen.

Of course, these changes make no difference to the Big River. It will continue to rise and fall, erode and deposit, shape the land and history in its own way, as all big rivers do. However, the upper Yukon, the part that is preserved, does not run entirely free anymore. It is managed wil-derness now, if such has any meaning. It may continue to shape hearts, minds, and souls of men and women who are willing to let it; who will work a little harder at it, seeking out the quiet corners; who will chose to go softly and engage the land on its own terms with minimal trap-pings. There's still plenty of room to get lost out there so as to find one-self. But a permit may be required.

NOTES

Chapter I

1. This changed with the construction of the North Slope Haul Road in 1974, a hastily built supply road for the construction and maintenance of the Trans-Alaska Pipeline. The road was renamed the Dalton Highway and opened to the public (with permit) to mile 211 in 1981. In 1994 the entire four hundred fourteen miles were opened without permits.

2. Patty, 1971. 2–18.

3. Cooke, 1964. The idea for the Rampart Dam Project first surfaced in the 1950s, and ten years later, in spite of its potential impact on the culture and environment of the entire Yukon Flats area, its economic merits were still being argued. See, for example, Spurr et al., 1966, and Schramm, 1968.

4. Lautaret, 1989a. 133.

5. Lautaret, 1989b. 135.

6. Stuck, 1917. 88.

Chapter II

1. Hudson Stuck tells of meeting a pole-boater in Circle who had just completed a journey of six hundred miles from the Koyukuk River downstream, poling his long, narrow boat twenty miles a day on average against the Yukon current. The man commented that he didn't mind it much, except in the Flats, following with this quote. Stuck (1917). 64.

2. Murray, 1910.

3. Stuck, 1917. 40.

4. Schwatka, 1885. 134.

5. Kutchin is synonymous with Gwich'in in present day usage. The name

Kutchin ("one who dwells"), usually preceded with a geographical designator, was used in earlier literature in reference to the Athabaskan tribes occupying the area west of the Makenzie Delta, generally in the Peel, Porcupine, and middle Yukon drainages. "Kutcha," an English transliteration, was the name Cornelius Osgood was given by Chandelar, Crow, and Peel River informants for the Yukon Flats people, which together with "Kutchin" meant "those who dwell on the flats." Hence, "Yukon Flats Kutchin" was also frequently used in early literature (see Osgood, 1934). Currently the Gwich'in International Council (2009) recognizes the native Athabaskans of Fort Yukon as Gwich'yaa Gwitch'in, and those from Circle, Danzhit Hanlaih Gwich'in.

6. Slobodin, 1981.

7. McClellan and Denniston, 1981.

8. Nelson, 1983.

9. According to the Alaska Native Language Center (University of Alaska, Fairbanks) there are presently about three hundred speakers of the Kutchin (Gwich'in) language in Alaska, while the Han of Eagle and Dawson number about fifty people, with only twelve still speaking the native language. In Canada, the Yukon Native Language Center writes: "In Dawson City there is only a handful of fluent [Han] speakers remaining. The rapid decline of the language in this region is due in large part to the dramatic changes brought by the flood of outsiders with the Gold Rush of 1898. There are more speakers in Eagle and Fairbanks, Alaska, but probably fewer than fifteen."

10. Murray, 1910. Russia first offered to sell Alaska to the United States in 1859, but it was not until after the Civil War that Secretary of State William Seward took up the proposal. Alaska was officially transferred to the United States on October 18, 1867, under President Andrew Johnson.

11. Schwatka, 1885. 11.

12. Whymper, 1868.

13. Schwatka, 1885. 281.

14. Andrews, C. L. 1941,. 200 cited by: Sherwood, M. B. 1965, 147. The quote is attributed to Stewart Menzies, most likely an Alaska Commercial Company employee.

15. The two prospectors were identified as Indians by Richard Mathews (1968) and their names given as Syroska and Pitka." Melody Webb (1985. 89) identified them as Sergi Gologoff Cherosky and his brother-in-law Pitka Pavaloff, and thereafter referred to them as Cherosky and Pitka. An earlier source, however, identifies them as "Russian half-breeds Poitka and Sonoiska" (Goodrich. 118). Hudson Stuck (1917. 86) mentions the "Russian half-breeds Sarosky and Pitka" and claims to know Pitka, whom he says William Dall had brought up from Nulato to cook for him in 1867.

16. Here again, accounts differ. According to Mathews, the population "swelled to over 1000" by 1896. Stuck, however, put it at three thousand.

17. Stuck, 1914. 369.

18. Falk, 1991. 27-34.

19. Stuck (1917. 113) appears to be the source of this information, though he did not assign a specific date other than to suggest that the fish wheel "came as an incident of the stampede to Fairbanks." The year 1904 was given by Osgood (1971. 139) and apparently picked up by Caulfield (1979. 10).

20. Statistics on the operation of the dredge are from: United States National Park Service, Final Environmental Impact Statement, Volume I, Cumulative Impacts of Mining: Mining in Yukon-Charley Rivers National Preserve, Alaska, 1990. 37-38.

21. Ibid.. 37-38, 40.

Chapter III

1. The early French experience in Arkansas, including this visit by Father Poisson, is described in detail by Philip Marchand, 2005. Father Poisson's lack of experience with mosquitoes was evident in his noting (p.369) that the "cruel persecution" by the mosquitoes "passes all belief "and that "certainly nobody in France could imagine it."

2. Sherwood, 1965, citing William Dall, Alaska and its Resources. 92, 100.

3. Murray, 1910.

4. Schwatka, 1885.

5. An overview of the biota and climate of the Beringian region during the late glacial period in Alaska can be found in Young, 1976. 124-128.

6. Murray, 1910. 84.

7. Whymper, 1868. 223.

8. Crow and Obley, 1981.

9. Schwatka, 1885. 251-256.

10. Ibid.. 262.

11. McClellan and Denniston, 1981. 377.

12. Stuck, 1917. 84.

13. United States Senate Committee on Military Affairs, 1900. 499.

14. Ibid. 499.

15. Brooks, 1973. 346-346.

16. United States Senate Committee on Military Affairs, 1900. 500.

17. Postell, 1910. 38-39.

18. Stuck, 1917. 82.

19. Enacted by the Legislature of the Territory of Alaska on April 27, 1915. The law became obsolete in 1923 when Congress granted citizenship to any

Indian born within the territorial limits of the United States. See Lautaret, 1989. 88-89.

20. Nelson, 1983.

21. Grizzly bears are heaviest just before denning. According to the Alaska Department of Fish and Game (ADFG), at this time most mature males weigh between five hundred-nine hundred lbs (one hundred eighty-four hundred ten kg). (The Smithsonian puts the mean weight of male grizzly bears in Alaska at three hundred eighty nine kg.) Extremely large individuals, notes ADFG, can weigh as much as fourteen hundred lbs (six hundred forty kg). According to the same source, bull moose in prime condition generally weigh from twelve hundred-sixteen hundred lbs (five hundred forty two-seven hundred twenty five kg). The largest *confirmed* Alaskan moose was reportedly taken on the Yukon River in 1897 and weighed seventeen hundred ninety-nine lbs (eight hundred eighteen kg). The source of this often quoted record appears to be The Guinness Book of Animal Facts and Feats, Sterling Publishing Co., Inc., 1983.

22. James Huntington was born to an Athabaskan mother and white trapper, and spent his life in various villages along the Koyukuk and Yukon Rivers. An account of his life in interior Alaska was recorded by writer Lawrence Elliott and published as: Huntington, J. and L. Elliott. 1966. *On the Edge of Nowhere*. Crown Publishers, NY. A third edition was printed by Epicenter Press, 2002. In this most interesting read, Huntington talks at one point about bear hunting in the fall, about how his Uncle Hog River Johnny went after a bear with an ax, and about his own experience with the same ("All these things, and more, my Uncle Johnny taught me, and it was a rare fall that I didn't pack home plenty of bear meat.") In his discourse about den hunting, Huntington states: "...the natives have hunted autumn bears for generations, mostly with axes to save ammunition, and they really know the business." James Huntington was elected to the Alaska state legislature in 1974.

23. For an interesting summary of Rogers' early work, see: Rogers, L. 1981. A Bear in Its Lair. Natural History 90: 64-70.

24. Referenced by Osgood, 1971. 110.

25. Stuck, 1917. 82.

26. Amundsen, 1908. Amundsen did indeed have important news to send: With his sighting of the ship Charles Hanson on August 25, 1905, steaming towards the Gjoa from the Pacific, Amundsen knew he and his crew of six had, after two years at sea, finally succeeded in their quest: " The North West Passage had been accomplished — my dream from childhood. This very moment it was fulfilled." Interestingly though, his record of the moment is filled more with feelings of great nostalgia for his homeland than a sense of satisfaction

with his discovery. "My home and those dear to me there at once appeared to me as if stretching out their hands — " (126). As he approached the Fort Egbert telegraph station several months later, he again expressed these feelings, "You can imagine how over-powering is the thought that within a few hours you will be in touch with the dear ones at home." Amundsen stayed two months in Eagle awaiting news from Norway, and of these two months he wrote "I shall never forget that time, as it is associated with some of my most cherished and pleasant recollections." (246)

27. Ibid. 245.

28. Stuck, 1917. 82 seems to be the sole source of this story. O'Neill has kept the story alive (2006. 154), quoting Stuck as well, but apparently the woman's name was never recorded.

29. Ibid. 83

30. Marten population densities and individual ranges are estimated from data of Katnik et al., 1994, Thompson, 1994, Mowat and Paetkau, 2002, and Dumyahn and Zollner, 2007, and assume that 6% of a male's territory is shared with other males, 11% of a female's territory is shared with other females, and the average overlap between male territory and female territory is 54%.

31. Based on average fur prices from that period as recorded in Oregon Dept. of Fish and Wildlife records, 1999.

32. Schwatka, 1885. 265.

Chapter IV

1. The homestead act of 1862 was repealed in 1976, but with a ten-year extension to 1986 for eligible lands in the State of Alaska.

2. An entry area has since been added to the cabin. The story of "Phonograph" Nelson and later occupation of his cabin is related by historian Dan O'Neill (2006. 129), though a 25 year gap between 1950 and 1975 remains unaccounted for, during which time the cabin was apparently little used.

3. United States National Park Service, 1990. 41.

4. Ibid.. 44.

5. Brad Snow provided this description of ice-pan formation as he surmised it from his experience on the Nation and Yukon Rivers. Ice-pans also form from the agglomeration of frazil ice—small disc-shaped platelets of ice drifting in cold, running water—into large clusters. For an excellent description of river ice formation and breakup, see the New Brunswick River Ice Manual, 1989.

6. For a more detailed discussion and review of the literature on this subject, see Marchand, 2014. 168-171.

7. Evidence supporting the possible role of plant defense compounds in regulating hare population cycles has come from a number of interesting field tests, including elaborate reciprocal feeding trials with Alaskan, Siberian, and Finnish hares, utilizing shrubs with both high and low concentrations of defense compounds. For a summary of these studies, see Marchand, 2014. 159-168.

Chapter V

1. Stuck, 1917. 79.

Chapter VI

1. This story is told in considerable detail by John McPhee (1976), who took the trouble to track down and interview Leon Crane, then living in Philadelphia, some thirty years after the plane crash. A later account was published by Captain Stephen M. Morrisette (1990), filling in military details of the ill-fated flight and the events leading up to the pilot's order to bail, and adding entertaining anecdotes about Crane's eventual return to his base (then Ladd Field). While not specifically stated, Morrisette apparently interviewed Crane again for the story. At the time of writing, Crane was still living in Philadelphia.

2. In his research on the early homesteaders of the Yukon and Charley Rivers area, historian Dan O'Neill found Phil Berail "at the top of everyone's list of 'the toughest men I ever knew.'" According to O'Neill, Berail trapped and mined in the area for "thirty or forty years."

3. Dearing, 1997.

4. McKechnie et al., 1994. Pikas almost always rob from the opposite sex (with no gender bias among thieves), but this is likely related to the fact that pikas tend to establish territories next to the opposite sex and haypile thievery is almost always from immediate neighbors.

5. Lidicker and Batzli, 1999. 632-633.

6. Bergerud et al., 1984. 10. Provides population and range estimates for the Fortymile caribou herd from the 1920's to the 1970's, citing several sources.

7. Ibid.. 11, 19.

8. Caulfield, 1979. 38.

9. Ibid.. 31.

10. McEwan and Whitehead, 1970.

11. Mattson and Reinhart, 1997.

12. Smith, 1968.

13. Boertje, 1985.

14. In 1973, a second atmospheric monitoring station was established at Point Barrow, Alaska. Both stations, Mauna Loa and Point Barrow, have

recorded steadily rising CO_2 concentrations into the 21st century, reaching a landmark four hundred ppm in 2013. When monitoring began at Mauna Loa in 1958, atmospheric CO_2 was at three hundred seventeen ppm. See chapter 7 for the longer-term effects in Alaska of this trend.

15. Cook, 1989.

16. See, for example, papers by Cwynar and Ritchie, 1980; Edwards and Armbruster, 1989; Lloyd et al., 1994.

17. For a detailed discussion of the changes associated with freezing acclimation in plants, see Marchand, 2014. 48-59.

18. Besides facilitating the grinding of coarse material in the gizzard, ingestion of grit provides a number of essential elements and trace minerals. However, it may also result in exposure to toxic elements such as cadmium where they are present in soils and groundwater. An interesting chemical analysis of crop and gizzard (crop plus grit) contents is provided by Bendell-Young and Bendell, 1999.

19. A good literature review and recent empirical study of body mass changes and behavioral response to heat stress in moose is found in van Beest and Milner, 2013. See also Lenarz et al., 2009.

20. The literature regarding wolf-moose interactions is extensive, but an excellent and accessible overview based on fifty years of study at Isle Royale is provided by Vucetich and Peterson, 2012. A comprehensive discussion of the winter biology of moose is provided by Marchand, 2014. 189-194.

Chapter VII

1. Numbers can vary widely from year to year. In a letter from Willard Grinnell (a contemporary of Morris and Silas Gundrum) to Dave Evans, reported by O'Neill (2006. 163), Grinnell credited the Gundrums with trapping one hundred seventy five marten between them in one winter.

2. Marchand, 1976.

3. A good summary of temperature change in Alaska from 1949 to 2005 can be found at http://climate.gi.alaska.edu/ClimTrends/Change/TempChange.html.

4. The literature is voluminous, but for a good entry see Tape, et al., 2006 (shrub expansion) and Harsch et al., 2009 (treeline advance).

5. See, for example, Wilmking et al., 2004.

6. Barber et al., 2000, provides tree-ring evidence for increasing drought stress in interior Alaska as a result of climate warming.

7. O'Neill, 2006. 182 and 228, states that the 1991 Charley River fire was started by flares from dogfighting jets overhead. This information seems well-guarded, as I could find no confirmation from government sources.

8. History and acreage of burns comes primarily from Sorbel and Allen, 2005 (which includes only those fires for which burn severity maps have been generated using Landsat imagery), and the National Park Service Fire and Aviation Management Site, http://www.nps.gov/applications/fire/wildland-fire/fires/reports-and-archives.

9. Barnes, 2013.

10. O'Neill, 2006.

11. Visitation figures by month and year are from the NPS Stats Report Viewer, https://irma.nps.gov/Stats/SSRSReports/Park%20Specific%20Reports/All%20Recreation%20Visitors%20By%20Month?Park=YUCH

12. Morrisette, 1990.

REFERENCES

Ager, T. 1975. Late Quaternary environmental history of the Tanana Valley, Alaska. Ohio State University Institute of Polar Studies Report No. 54. 117.

Andrews, C. L. 1941. Some Notes on the Yukon by Stewart Menzies. Pacific Northwest Quarterly 32:200.

Amundsen, R. 1908. The North West Passage: Being the record of a voyage of exploration of the ship "Gjöa", 1903-1907, by Roald Amundsen with a supplement by First Lieutenant Hansen, Vice-Commander of the expedition. Vol. II. Archibald Constable and Company Limited, London.

Barber, V. A., G. P. Juday, and B. P. Finney. 2000. Reduced growth of Alaskan white spruce in the twentieth century from temperature induced drought stress. Nature 405(6787):668-673.

Barnes, J. 2013. Shortened fire return intervals in Alaska boreal forests. April 4, 2013 presentation to Alaska and Northwest Wildlife Society. Unpublished.

Bendell-Young, L. I. and J. F. Bendell. 1999. Grit ingestion as a source of metal exposure in the spruce grouse, *Dendragapus canadensis*. Environmental Pollution 106:405-412.

Bergerud, A. T., R. D. Jakimchuk, and D. R. Carruthers. 1984. The buffalo of the North: Caribou (*Rangifer tarandus*) and human developments. Arctic 37(1):7-22.

Boertje, R. D. 1985. An energy model for adult female caribou of the Denali herd, Alaska. Journal of Range Management 38:468-473.

Brooks, A. H. 1973. Blazing Alaska's Trails. University of Alaska Press, Fairbanks, 2nd edition.

Caulfield, R.A. 1979. Subsistence use in and around the proposed Yukon-Charley National Rivers. Occasional Paper No. 20, Anthropology and Historic Preservation Cooperative Park Studies Unit, University of Alaska, Fairbanks.

Cook, Francis H. 1989. The Jewel Net of Indra. In: Callicott, J. Baird and Roger T. Ames (eds), Nature in Asian Traditions of Thought: Essays in Environmental Philosophy. State University of New York Press, Albany. 213-229.

Cooke, A. 1964. The Rampart Dam proposal for the Yukon River. Polar Record 12: 277-280.

Crow, J. R. and P. R. Obley. 1981. Han. In: Helm, J. (ed.), Handbook of North American Indians, Vol. 6, Subarctic, Smithsonian Institute, Washington, D.C.

Cwynar, L. C. and J. C. Ritchie. 1980. Arctic steppe tundra: A Yukon perspective. Science 208:1375-1377.

Dearing, M. D. 1997. The manipulation of plant toxins by a food-hoarding herbivore, *Ochotona princeps*. Ecology 78(3):774-781.

Dumyahn, J. B. and P. A. Zollner. 2007. Winter home-range characteristics of American marten (*Martes americana*) in northern Wisconsin. Amer. Mid. Naturalist 158(2):382-394.

Edwards, M. E. and W. S. Armbruster. 1989. A tundra-steppe transition on Kathul Mountain, Alaska, USA. Arctic and Alpine Research 21(3):296-304.

Falk, T. 1991. Urban turnaround in Sweden: the acceleration of population dispersal, 1970-1975. GeoJournal 2(1):27-34.

Goodrich, H.B. History and Conditions of Yukon Gold District to 1897. US Geological Survey, 18ᵗʰ Annual Report, Part III, Washington, D.C.

Harsch, M. A., P. E. Hulm, M. S. McGlone, and R. P. Duncan, "Are Treelines Advancing? A Global Meta-Analysis of Treeline Response to Climate Warming," Ecology Letters 12 (2009):1040-1049.

Huntington, J. and L. Elliott. 1966. On the Edge of Nowhere. Crown Publishers, NY.

Katnik, D.D, D. J. Harrison, and T.P. Hodgman. 1994. Spatial relations in a harvested population of marten in Maine. J. Wildlife Management 58(4):600-607.

Lautaret, R. L., 1989a. Alaskan Historical Documents since 1867. 34. Project Chariot. McFarland, Jefferson, NC. 133.

Lautaret, R. L., 1989b. Alaskan Historical Documents since 1867. 35. Point Barrow Conference on Native Rights. McFarland, Jefferson, NC. 135.

Lenarz, M. S., M. E. Nelson, M. W. Schrage, and A. J. Edwards. 2009. Temperature Mediated Moose Survival in Northeastern Minnesota. Journal of Wildlife Management 73(4):503-510.

Lidicker, W.Z. Jr. and G.O. Batzli. 1999. Singing vole. In: Wilson, D.E. and S. Ruff, Eds., Smithsonian Book of North American Mammals. Smithsonian Institution Press, Washington.

Lloyd, A. H., W. S. Armbruster, and M. E. Edwards. 1994. Ecology of a steppe-tundra gradient in interior Alaska. Journal of Vegetation Science 5(6):897-912.

Marchand, P. E. 2005. Ghost Empire: How the French Almost Conquered North America. McClelland and Stewart, Ltd, Toronto.

Marchand, P. J. 1976. Growth and population structure of white spruce in the forest-tundra ecotone, Twin Mountain area, Alaska. In: Young, S. B. (ed.) The Environment of the Yukon-Charley Rivers Area, Alaska, Contributions from The Center for Northern Studies No. 9, Wolcott, VT.

Marchand, P. J. 2014. Life in the Cold, 4th edition. University Press of New England, Hanover, NH.

Mathews, R. 1968. The Yukon. Holt, Rinehart and Winston, New York.

Mattson, D. J. and D.P. Reinhart. 1997. Excavation of red squirrel middens by grizzly bears in the whitebark pine zone. Journal of Applied Ecology 34:926-940.

McClellan, C. and G. Denniston. 1981. Environment and Culture in the Cordillera. In: Helm, J. (ed.), Handbook of North American Indians, Vol. 6, Subarctic, Smithsonian Institute, Washington, D.C., 372-386.

McEwan, E. H. and P. E. Whitehead. 1970. Seasonal changes in the energy and nitrogen intake in Reindeer and Caribou. Canadian Journal of Zoology 48:905-913.

McKechnie, A. M., A. T. Smith, and M. M. Peacock. 1994. Kleptoparasitism in pikas (*Ochotona princeps*): Theft of hay. Journal of Mammalogy 75(2):488-491.

McPhee, J. 1976. Coming into the Country. Farrar, Straus, and Giroux, New York.

Morrisette, S. M. 1990. The man who walked out of Charley River. Flying Safety 46(7):15-19.

Mowat G. and D. Paetkau. 2002. Estimating marten (*Martes americana*) population size using hair capture and genetic tagging. Wildlife Biology 8(3):201-209.

Murray, Alexander Hunter. 1910. Journal of the Yukon, 1847-1848. Publication of the Canadian Archives, No. 4, Ottawa.

Nelson, Richard K. 1983. Make Prayers to the Raven: A Koyukon View of the Northern Forest. Univ. of Chicago Press, Chicago.

New Brunswick River Ice Manual, August 1989. Prepared by: The New Brunswick Subcommittee on River Ice, Environment Canada New Brunswick, Inland Waters Directorate, Department of the Environment.

O'Neill, D. 2006. A Land Gone Lonesome. Counterpoint, New York.

Oregon Dept. of Fish and Wildlife, Trapping White Paper, December, 1999

Osgood, Cornelius. 1934. Kutchin tribal distribution and synonymy. American Anthropologist, New Series 36(2):168-179.

Osgood, Cornelius. 1971. The Han Indians: A Compilation of Ethnographic and Historical Data on the Alaskan-Yukon Boundary. Yale University Publications in Anthropology No. 74, New Haven.

Patty, S.H. 1971. A Conference with the Tanana Chiefs. Alaska Journal 2:2-18.

Postell, A. 1910. Where Did the Reindeer Come From? Alaska Experience, the First Fifty Years. Amaknak Press, Portland, Oregon.

Rogers, L. 1981. A Bear in Its Lair. Natural History 90: 64-70.

Schramm, G. 1968. The economics of an upper Yukon basin power development scheme. Annals of Regional Science 2(1): 214-228.

Schwatka, Frederick. 1885. Alaska's Great River: A Popular Account of the Travels of the Alaska Exploring Expedition of 1883, Along the Great Yukon River From Its Source to Its Mouth, in the British North-West Territory, and in the Territory of Alaska. Cassell and Company, Ltd., New York.

Sherwood, M. B. 1965. Exploration of Alaska, 1865-1900. Yale University Press, New Haven. 147.

Slobodin, R. 1981. Kutchin. In: Helm, J. (ed.), Handbook of North American Indians, Vol. 6, Subarctic, Smithsonian Institute, Washington, D.C. 514-532.

Smith, M. C. 1968. Red squirrel responses to spruce cone failure in interior Alaska. Journal of Wildlife Management 32: 305-317.

Sorbel, B. and J. Allen. 2005. Space-based burn severity mapping in Alaska's national parks. Alaska Park Science 4(1): 5-11.

Spurr, S. H., E. F. Brater, and M. F. Brewer. 1966. Rampart Dam and the economic development of Alaska. Vol. 1, Summary Report, Rampart Dam-Alaska Economic Development Project, School of Natural Resources, University of Michigan, Ann Arbor.

Stuck, Hudson. 1914. Ten Thousand Miles with a Dog Sled: A Narrative of Winter Travel in Interior Alaska. Reprinted by Wolf Publishing Co., Prescott, Arizona, 1988.

Stuck, Hudson. 1917. Voyages on the Yukon and Its Tributaries: A Narrative of Summer Travel in the Interior of Alaska. Charles Scribner's Sons, New York.

Tape, K., M. Sturm, and C. Racine, "The Evidence for Shrub Expansion in Northern Alaska and the Pan-Arctic," Global Change Biology 12 (2006):686-702.

Thompson, I. D. 1994. Marten populations in uncut and logged boreal forests in Ontario. J. Wildlife Management 58(2):272-280.

United States National Park Service, Final Environmental Impact Statement, Volume I, Cumulative Impacts of Mining: Mining in Yukon-Charley Rivers National Preserve, Alaska, 1990.

United States Senate Committee on Military Affairs. 1900. Compilation of narratives of explorations in Alaska. Government Printing Office, Washington. Also Issued as Senate Report 1023, 56th Congress, 1st session.

van Beest, F. M. and J. M. Milner. 2013. Behavioral responses to thermal conditions affect seasonal mass change in a heat-sensitive northern ungulate. PLoS ONE 8(6):e65972. doi:10.1371/journal.pone.0065972.

Vucetich, J. A. and R. O. Peterson. 2012. The population biology of Isle Royale wolves and moose: an overview. URL:www.isleroyalewolf.org.

Webb, M. 1985. The Last Frontier. University of New Mexico Press, Albuquerque.

Whymper, Frederick. 1868. Travel and Adventure in the Territory of Alaska, Formerly Russian America—Now Ceded to the United States—and in Various Other Parts of the North Pacific. John Murray, London.

Wilmking, M., G. P. Juday, V. A. Barber, and H. J. Zald. 2004. Recent climate warming forces contrasting growth responses of white spruce at treeline in Alaska through temperature thresholds. Global Change Biology 10, 1724–1736, doi: 10.1111/j.1365-2486.2004.00826.x.

Young, S.B. 1976. Floristic investigations in the "arctic-steppe" biome. In: Young, S. B. (ed.) The Environment of the Yukon-Charley Rivers Area, Alaska, Contributions from The Center for Northern Studies No. 9, Wolcott, VT.

INDEX